THE GENETICS OF MENTAL RETARDATION

THE GENETICS OF MENTAL RETARDATION

Biomedical, Psychosocial and Ethical Issues

Edited by

E. K. HICKS

Netherlands Universities Joint Social Research Centre, Amsterdam, The Netherlands

J. M. BERG

Depts of Psychiatry and Genetics, University of Toronto and Director of Genetic Services and Biomedical Research, Toronto, Canada

Kluwer Academic Publishers

Dordrecht / Boston / London

Library of Congress Cataloging in Publication Data

The Genetics of mental retardation.

Papers presented at the 3rd international workshop of the
Bishop Bekkers Foundation and the Bishop Bekkers Institute held
April 22-24, 1986 in Doorn, The Netherlands.
Excludes index.
1. Mental retardation—Genetic aspects—Congresses. 2.
Mental retardation—Diagnosis—Congresses. 3. Prenatal
diagnosis—Congresses. 4. Genetic counseling—Congresses. I.
Hicks, E. K. II. Berg, J. M. (Joseph Maurice) III. Stichting
Bisschop Bekkers. IV. Stichting Bisschop Bekkers Instituut.
RJ506.M4G46   1988   616.85'88042   88-659
ISBN-13:978-94-010-7095-9   e-ISBN-13:978-94-009-1339-4
DOI: 10.1007/978-94-009-1339-4

Published by Kluwer Academic Publishers,
P.O. Box 17, 3300 AA Dordrecht, The Netherlands.

Kluwer Academic Publishers incorporates
the publishing programmes of
D. Reidel, Martinus Nijhoff, Dr W. Junk and MTP Press.

Sold and distributed in the U.S.A. and Canada
by Kluwer Academic Publishers,
101 Philip Drive, Norwell, MA 02061, U.S.A.

In all other countries, sold and distributed
by Kluwer Academic Publishers Group,
P.O. Box 322, 3300 AH Dordrecht, The Netherlands.

TABLE OF CONTENTS

SECTION III: GENETIC COUNSELLING

ACKNOWLEDGEMENTS

This volume is the product of the 3rd International Workshop of the Bishop Bekkers Foundation and the Bishop Bekkers Institute (B.B.I). The workshop, sponsored by the foundation, was held on 22-24 April, 1986, in Doorn, The Netherlands, at the Bartimeushage Centre. The Bishop Bekkers Foundation has a twofold aim. It is actively engaged in fund raising and support of activities concerned with the care of the mentally retarded in the Netherlands, as well as fostering promising international initiatives in the area of mental retardation by providing organisational, public relations and financial support to such endeavours.

The Foundation derives its name from the late Roman Catholic Bishop, Mgr. W. M. Bekkers, who was actively concerned for the underprivileged in general, and the mentally retarded in particular.

In 1972, the Foundation, long aware of the paucity of research efforts concerned with mental retardation, established the Bishop Bekkers Institute (B.B.I.), which it, together with the Dutch government, continues to subsidize. This Institute promotes research on mental retardation and mental retardation services by stimulating new research projects, publishing a scientific quarterly, and providing library services.

The choice of the subject matter for the 3rd International Workshop was the result of consultations at the 7th Congress of the International Association for the Scientific Study of Mental Deficiency, held during March 1985 in New Delhi. There was a clearly felt need for up-to-date information on medical, psychosocial and ethical implications of developments in the field of prenatal and postnatal diagnosis. To meet this need, an international workshop of specialists was deemed necessary.

We express our appreciation to the organisors and participants of this workshop. We are especially indebted to the workshop chairman, Professor Ignacy Wald, who devised the programme and selected the participants. We also extend our appreciation to the editors of this volume. Dr. E. K. Hicks transcribed and summarised the discussion sections, and prepared the introduction and the final draft of the papers. The appearance of this volume is due to her tireless and enthusiastic work. Professor J. M. Berg and Professor M. F. Niermeijer provided an intensive knowledge of recent developments in the field of genetics in their insightful editing of the papers, the introduction and the discussion summaries. We also express our thanks to the Vereniging Bartimeus in Doorn, and the director of its Centre for the Multiple Handicapped (Bartimeushage), Dr. C. G. A. de Jong, for the extensive hospitality shown to the workshop participants by his staff. A final acknowledgement is to Dr. Annalise Dupont, President of the International Association for the Scientific Study of Mental Deficiency, for her goodwill and active support.

Dr. A.L.M. Knaapen
Chairman, Bishop Bekkers
Foundation

Mr. B. van Zijderveld
Member of the Board
Bishop Bekkers Foundation

G.J.M Beeker, MD
Chairman, Bishop Bekkers
Institute

Dr. S. M. Nemeth
Director, Bishop Bekkers
Institute

EDITORIAL PREFACE

The Bishop Bekkers Foundation, devoted to the welfare of those with mental handicap and to the amelioration and prevention of this and related disabilities, is to be warmly congratulated for sponsoring and organizing the 1986 International Workshop from which the present book is derived. With commendable foresight, the Foundation recognized that genetic aspects of mental handicap were a timely focus for the Workshop and that dramatic *biomedical* developments and prospects in this sphere have highly significant *psychosocial* and *ethical* ramifications.

The papers of the 23 contributors in each of these areas, together with an introductory essay and discussion summaries, comprise this volume. Much of the subject matter is inevitably concerned with such sensitive issues as sanctity and quality of life and the – sometimes contrasting and even in conflict – rights and needs of the actually and potentially handicapped, their relatives and society in general. Such considerations, not surprisingly, engender different approaches and viewpoints. The papers in this book reflect this, with the editing attempting to achieve, as far as possible, a coherent and consistent format and not an identical outlook. The perceptions and views expressed in each paper are therefore those of the author concerned and not necessarily those of the editors or of the Bishop Bekkers Foundation. This is, of course, as it should be, and it is anticipated that different viewpoints expressed by different authors will be a useful and constructive contribution to the important continuing debate as to what is reasonable and proper with respect to the rapidly growing technical developments discussed in the text.

As editors, we thank the authors for their co-operation and, as participants in the Workshop, we express our appreciation, on behalf of all the other participants also, for the generous support of the Bishop Bekkers Foundation and for the gracious hospitality of the Foundation and of our hosts at the Workshop site (Bartimeushage in Doorn).

The Editors

INTRODUCTORY STATEMENT

During recent decades there has been a considerable increase in knowledge concerning the role of genetic factors in mental retardation with respect to monogenic, multifactorial and chromosomal disorders. Many new genetically determined diseases leading to mental retardation have been recognised, and the potential for diagnosis, prevention, and even treatment has increased immensely.

Some of the more important developments in genetic research, relevant to mental retardation, include:

– Increased knowledge about the heterogeneity of mental retardation. Nosological research has split seemingly single disorders into a number of independent entities. Biochemical research has similarly expanded, significantly increasing the list of types and variants of biochemical disorders.

– Considerable expansion of the role of cytogenetic analysis in studying mental retardation. The identification of fragile sites (particularly fragile X syndrome) exemplifies this progress. Also the development of microcytogenetics has facilitated research on subtle chromosomal aberrations associated with various conditions (including some monogenic disorders).

– DNA analysis, using restriction enzymes. This not only allows the precise mapping of genes, but is also invaluable in the early diagnosis of many genetic disorders (even those in which the primary gene defect is unknown, e.g. Huntington chorea).

– Prenatal diagnostic advances. Techniques are now available for the recognition of chromosomal anomalies, some malformations, and some biochemical disorders. The supplementation of amniocentesis with various visualisation techniques and trophoblast examination has added new possibilities for prenatal diagnosis. There is increasing recognition of the future potential for treating some of the genetic disorders, even in a fetus.

Progress in genetic research has also changed our understanding of genetic factors in mental retardation. For example, traditionally, severe retardation was assumed to be mono- factorial, while mild retardation was held to be multi-factorial. The recognition of the fragile X syndrome helped to change this simplistic model. Improved understanding has also considerably increased the possibility of prevention.

Given these technical developments, the need for increased consideration of psycho-social and ethical issues is readily apparent.

The purpose of the present workshop on the genetics of mental retardation was to discuss biomedical advances and their psychosocial and ethical conseqences. The Bishop Bekkers Foundation brought together an international and interdisciplinary group of specialists, in the fields of genetics, pediatrics, psychology and ethics, to address these considerations from a cross-cultural perspective.

Ignacy Wald   XI

INTRODUCTION

The present volume comprises a compilation of papers and discussion summaries from the 1986 Bishop Bekkers – sponsored workshop on 'The Genetics of Mental Retardation: Medical, Psycho- social and Ethical Issues.' A primary concern of both the workshop organisers and participants was the increase in ethical, moral and legal controversies associated with the current array of choices provided by clinical genetics.

As indicated in the introductory statement by Professor Wald, even a cursory review of the biomedical literature published during the past decade demonstrates the increasing extent and sophistication of knowledge about the role of genetic factors in mental retardation. This has greatly expanded the potential for diagnosis, prevention and treatment of genetic disorders associated with such retardation. These technical advances and their complex social implications have necessitated a re- evaluation of many basic moral and ethical values.

One major ethical/legal controversy relates to the perception that there is an inherent and absolute value to human life, which must be preserved in any medically-related decision-making process. With advanced biomedical technology, however, this perception is no longer tenable without qualification. Specifically, there is no longer concern only with the preservation of human life, but also with the actual or potential 'quality of life' (implying that there is more to life than merely being alive). Clearly, however, the problem does not end there. How should 'quality of life' be evaluated?; when and by whom should a decision be made that an affected individual or fetus will have an unduly impoverished quality of life?; at what point should such a judgement be translated into an explicit decision to terminate or continue a pregnancy? Such ethical-moral issues are not only the subject of philosophical debate, but are connected with juridical systems in which decisions are often based upon the historical traditions of a given population.

Another important area of concern, where an individual (or fetus) is afflicted with mental retardation, is the family, which will have to provide the support system necessary to the life of such an individual. Such a support system involves:
– the daily care of the individual, the extent of which is related to the degree of mental retardation;
– allocation of the necessary time and funding where medical or other special treatment is required;
– integrating the afflicted individual into the family sphere, both immediate and extended, such that interrelationships between family members (especially when there are other children in the family) do not suffer;

– the organisation of the individual's life external to the home, in order to expand that person's potential for interaction within the broader social milieu.

It is the individual family which, as a unit, must determine its capability of coping with a mentally handicapped family member permanently in its care. Here also, ethics is involved – e.g. to what extent, if any, should the physician or geneticist be directive (i.e. actively involved in any family decision-making process)?; should the geneticist be selective, on a case-by-case basis, of the information imparted to the individual or couple relative to such issues as paternity, risks and carrier status, or should all available information be communicated?

These and related issues are reflected in both the papers and the discussion summaries included in this volume.

The volume is divided into three main areas: Prenatal Diagnosis, Postnatal Diagnosis and Genetic Counselling. This is done in the interests of organizational convenience and clarity, and does not imply that these areas are considered to be mutually exclusive – as, indeed, they are not.

Section I provides an overview of the advances in genetic research in prenatal screening and diagnosis, and the psycho- social and ethical issues resulting from the expansion of knowledge and available choices. Also considered are the uncertainties involved in some types of prenatal diagnosis, and the concomitant ethical issues raised by such uncertainty.

In the first paper in this volume, M. A. Ferguson-Smith stresses the need for expansion of prenatal screening services for at-risk women. Such services, in tandem with prenatal diagnosis and as an adjunct to genetic counselling, not only lighten the burden of genetic disorders at the community level, but also open more and earlier options for couples at risk of having an affected offspring. The widespread utilisation of prenatal spina bifida screening programmes attests to the acceptability of prenatal diagnosis by both the public and the medical profession. He summarises and evaluates currently available techniques which facilitate the prenatal diagnosis of many genetic disorders, outlines the medical indications for their use, and considers the impact that well-organised and widely available prenatal screening and diagnosis could have on the birth incidence of children with chromosome aberrations, metabolic disorders and neural tube defects.

Brambati, a pioneer in the development of first trimester fetal diagnosis by means of chorionic villus sampling (CVS), compares this method with more traditional amniocentesis. He considers the differences and risk factors involved in both methods, and stresses the advantages of CVS for a couple for whom termination of pregnancy is deemed advisable. Because CVS can be accomplished in the first trimester of pregnancy, in contrast to second trimester amniocentesis, it reduces psychological stresses for a couple, affords greater privacy, and results in fewer social and medical complications.

Mikkelsen presents preliminary results of a collaborative study, indicating the range of detectable cytogenetic aberrations using CVS. Although the data generally corroborate the psychological and medical advantages of CVS, she discusses the need for international cooperation in researching the risk factors and reliability of the procedure.

While M. E. Ferguson-Smith agrees that these new testing methods are valuable in the practice of preventive medicine, she points out that their accuracy is dependent on the efficiency and competence of available laboratory staff and facilities. She stresses the need for appropriate guidelines on the standards expected from a laboratory, and comments on what such standards should be.

Zwinger provides data on relatively extensive experiences with fetoscopy at a centre in Czechoslovakia, and discusses developments there concerning CVS.

Although the technical advances in prenatal diagnostic capability have increased the options for at-risk couples, they have also evoked considerable moral-ethical debate. Furthermore, they have increased the psycho-social dilemmas faced by at-risk potential parents and their clinical and counselling advisors. Thomassen-Brepols considers the psycho-social ramifications of prenatal diagnosis, arguing that, while this technical capability frequently provides early relief of parental anxiety, it often amplifies the stress and anxiety when a decision involving the termination of a pregnancy has to be considered. This is compounded by a general ethical uncertainty, and by the fact that many who find themselves faced with such a decision live in a social environment which morally and legally prohibits euthanasia.

Gustavson emphasizes the responsibility of the clinician to help at-risk couples couples realise that, when they opt for diagnostic procedures, they must also carry the decision-making responsibility when test results are positive. Wilken, on the other hand, does not seem to support parental decision-making autonomy, arguing that pregnancy termination is 'therapeutic' rather than 'preventive,' and should therefore be strictly controlled.

From a broader perspective, Fletcher traces the evolution of ethical issues associated with prenatal diagnosis. He argues that, in many societies, ethical issues evolve in two stages. In the first stage prenatal diagnosis is challenged by those concerned with the protection of the fetus and the handicapped. The second stage involves the three areas of parental decision-making autonomy, a beneficence approach in advising individuals and family members, and voluntary prenatal screening programmes. He stresses the need for concise ethical guidelines, pointing out that with mainly an oral ethical tradition at our disposal and the rapid rate of technological development, the ethical issues of the future will be sufficiently complex that we risk the codification of medical guidelines at the governmental, rather than the medical/professional, level.

In Section II postnatal diagnostic and ethical issues are considered. As in the prenatal period, it is deemed that, while both screening and other diagnostic testing should be made generally available, utilisation should be on a voluntary basis. This has the consequence that, with the exception of localities where screening programmes are mandated, genetic abnormalities which can currently be recognized may go undetected. Warburton discusses the organisational and socio-ethical problems associated with the mandated New York State newborn screening programme for haemoglobin abnormalities. In this case, there are serious problems in informing and obtaining access to patients for follow- up study and counselling. She uses the example of sickle cell anaemia to illustrate social-ethical concerns which can arise as screening capabilities expand.

The accurate diagnosis of dysmorphic syndromes is of obvious importance in genetic counselling. Baraitser illustrates the prowess of computer data base utilisation as a diagnostic aid in identifying cases involving multiple malformations and normal chromosomes. There are more than 1,000 recognisable non-chromosomal malformation syndromes. Given the number and variablility of these syndromes, a readily accessible cross- referencing catalogue of manifestations is invaluable. He illustrates the mechanics of his computer reference catalogue and its uses.

The papers by Verma, Niermeijer, Smithells and Heiberg deal with many of the diagnostic and counselling problems associated with mental retardation. Verma reviews the epidemiology of mental retardation in India and discusses teratogenic and consanguineous aspects of mental retardation in that country. These data are of considerable relevance for all developing ocuntries. Niermeijer considers some of the major technical difficulties involved in diagnosing actual or potential genetic abnormalities, and reviews the diagnostic strategies currently available.

Smithells stresses the importance of certainty in the diagnosis of an abnormality and in the determination of the severity of mental retardation in a newborn. He comments on the post-diagnostic problems encountered by both physician and parents where decisions must be made about intervention or non-intervention in the life of a seriously handicapped infant, and the consequences of each decision.

Heiberg describes the diagnostic and counselling problems arising from the heterogeneous aetiology of mental retardation. He outlines a Norwegian approach to the diagnosis and counselling of the mentally handicapped and their parents. The Frambu health centre programme is an example of the manner in which such institutions disseminate relevant health and welfare information to affected individuals and their families.

Postnatal diagnosis also involves ethical issues. Dunstan outlines the choices available to the clinician in the event of the birth of a severely handicapped child. Clinical management should be consistent with morality, professional ethics and the law. The principles relating to the protection, prolongation and taking of life are examined, with their application to the work of medical geneticists. Neither the moralist nor the parents may dictate the choice of clinical action; that responsibility, after consultation, lies with the medical practitioner, within the authority vested in him by the community.

Some major difficulties, both medical and ethical, of effective genetic counselling for mental retardation are considered in Section III. Here again there is some disparity between the rate of biomedical advances and the ability to determine what constitute appropriate reproductive decisions within the framework of existing moral-ethical standards. Moreover, since these standards often do not condone pregnancy termination, and remain silent on the subject of opting for prenatal (or other) diagnosis, it is not surprising that, while public opinion generally favours screening and diagnostic programmes, members of that public may find themselves in a moral-ethical dilemma when test results indicate fetal abnormality, for which termination of pregnancy is an option.

Berg, whose paper opens Section III, indicates that, although genetical counselling in mental retardation has attained a considerable level of sophistication, aetiological heterogeneity and variability in phenotypic outcome can make accurate and effective counselling difficult. He discusses the biomedical risk factors responsible for mental retardation in various circumstances, and emphasizes that the role and responsibilities of the counsellor extend beyond merely providing statistical information on risks of occurrence or recurrence.

Frijns illustrates the difficulties of effective counselling for mental retardation in relation to the fragile X syndrome. He points out that there is no exact correlation between this chromosomal abnormality and its variable phenotypic expression. Further complicating technical and, ultimately, counselling effectiveness is that cytogenetic screening procedures are less than ideal at present.

Dupont considers diagnostic and counselling problems in retarded children referred to a child psychiatric hospital. These problems range from difficulties in aetiological diagnosis to the application of effective counselling techniques when the presence of mental retardation is related to some psycho-social consideration. Collaboration between psychiatrists and geneticists may aid in differentiating between organic/genetic and psycho-social causes of retardation.

Czeizel describes a programme aimed at the prevention of mental retardation and the general improvement of the health of newborns in Hungary. The programme involves 'optimal' family planning, comprehensive medical and psychological examination of participating couples, corrective treatment where feasible, and limited follow-up procedures. He argues that the programme has decreased the incidence of genetic abnormalities and mental retardation.

Zaremba's discussion of tuberous sclerosis, and its phenotypic variability, provides an example of ethical issues in genetic counselling. Because a parent transmitting the condition may not be obviously affected, counselling raises numerous problems. Where information has no practical importance for prevention or treatment, and where such information is deemed to be potentially disruptive to the individual or family, nondisclosure of such information should be left to the discretion of the counsellor. Zaremba emphasizes that the primary consideration of genetic counselling must be the welfare of patients and their families.

De Wachter more directly focuses on ethical issues in genetic counselling, and proffers possible ethical approaches to such counselling.

The final paper, by S. and H. Kessler, deals with psychosocial aspects of genetic counselling. The emphasis here is on the cogitive problems arising in the communication between counsellor and those counselled. They stress that, although the counsellor often presumes that those seeking information have come with the intent of making reproductive decisions, this is not always the case. Decisions may have already been made subliminally prior to the counselling session. Consequently, the counsellor needs to develop communication techniques which can help persons counselled to effectively evaluate their reproductive decisions.

The papers in this volume consider both the range and consequences of biomedical advances in the diagnosis, treatment, and prevention of mental retardation and related abnormalities. Although the primary responsibility of the geneticist is the establishment of the relevant facts in an individual or family seeking counselling, and the communication of all that these imply to those concerned, the biomedical advances of the last decades have considerably expanded these responsibilities to include ethical and legal considerations in any decision-making process involving the reproductive options of at-risk persons. Consequently, the counsellor is not only responsible for the accuracy and communication of technical information, but also for the psycho-social consequences to which this information gives rise.

E. K. Hicks

SECTION I

PRENATAL DIAGNOSIS

RECENT DEVELOPMENTS IN PRENATAL DIAGNOSIS: BIOMEDICAL ASPECTS

The main purpose of medical genetics is to apply the principles of genetics to the practice of medicine so that the burden of genetic disorders can be reduced in the community. In practice, this consists first in identifying those members in the community who are at risk of developing a genetic condition or of having affected offspring. Those diagnosed accurately as being at risk can then be advised about the options open to them. This can be in the form of preventive therapy such as diet in the case of phenylketonuria, or surgical resection of the large bowel in the case of familial polyposis coli. Alternatively, if there is a risk of transmitting a serious disease to offspring, the reproductive options open to the person at risk can be explained so that informed decisions can be made. The process of providing this information and giving non-directive advice is known as genetic counselling.

Those at genetic risk in the community are frequently recognised only after a clinical diagnosis has been made in themselves or in a close relative. Thus the family at 1 in 4 risk of having a child with cystic fibrosis or muscular dystrophy will usually be oblivious of this tragedy until an affected child is born and accurately diagnosed. This occurrence may mean that the extended family may also be at risk, and so appropriate screening of close relatives may lead to the detection of other branches of the family equally at risk, but before the birth of an affected child.

The principle of genetic screening can be extended beyond the immediate family and used to detect genetic carriers in large populations at risk. For example, the high prevalence of Tay-Sachs disease in the Jewish community has led to successful screening programmes to detect carriers in North America, Israel and Western Europe. Similarly, the recent reduction of Beta-thalassaemia in Mediterranean countries including Northern Italy, Sardinia and Cyprus, has, in part, been due to successful carrier detection programmes.

Those couples at high risk of conceiving a child with a serious genetic disease have a number of effective options open to them. First, in a free society, they have the right to choose to go ahead and plan a pregnancy in the knowledge that the child may be affected. Secondly, they may choose any of a number of methods to avoid conception, including separation, contraception, and sterilisation. Thirdly, they may choose to have a healthy child by adoption, artificial insemination using donor semen (AID), or by ovum donation and *in vitro* fertilisation. Lastly, they may choose prenatal diagnosis and selective abortion if the fetus is found to be affected. It must be emphasised that parental decisions about these options can only be made satisfactorily if the couple are fully and honestly informed about the disorder and its prognosis. In genetic counselling there is no

3

E.K. Hicks and J.M. Berg (eds.), The Genetics of Mental Retardation, 3–14
© 1988 *by Kluwer Academic Publishers.*

place for concealing the implications of a diagnosis on the grounds that full revelation will upset the family, or create undue anxiety.

PRENATAL DIAGNOSIS

There is no question that the introduction of prenatal diagnosis and selective termination has been one of the most important adjuncts to genetic counselling. It is infinitely preferable to therapeutic abortion on the grounds of risk rather than of a firm diagnosis of fetal abnormality. It has allowed many thousands of women at high risk of serious fetal abnormality the reassurance to go ahead and plan for a healthy child. Many mothers have replaced the child lost due to a serious genetic disorder with a normal child, thanks to prenatal diagnosis. For most mothers the inevitable anxieties associated with waiting for a diagnosis in early pregnancy is much more preferable to the anxiety associated with the risk that a second child may be affected. In these mothers the natural abhorrence of an abortion is tempered by the knowledge that miscarriage is nature's way of dealing with the majority of developmental abnormalities. Thus, in over a half of all early miscarriages the fetus has a gross chromosome aberration, and in others the embryo is blighted or deformed. Selective termination following prenatal diagnosis is seen as the abortion of abnormal conceptions which somehow have escaped a natural process. Until primary prevention is possible, prenatal diagnosis is likely to continue as a widely used option for many serious disorders.

In discussions about selective termination it is frequently stated that the procedure is appropriate only for lethal, serious, handicapping disorders. Parents who have had experience of an affected child usually are in no doubt about what they regard as serious. Greater difficulty is encountered in prenatal screening programmes when a pregnancy is found to be affected in a couple who have no previous experience, and who thus require full information about the prognosis from their doctor. There is a temptation to make the decision for the couple, which must be resisted, for every couple should be allowed to make an informed decision in the light of their own feelings and wishes.

Medical geneticists rightly resist requests for prenatal diagnosis for trivial and non-medical indications. In most countries, termination of pregnancy for fetal indications that are not associated with a high risk of serious fetal abnormality would not be regarded as legal. This does not deter some couples from requesting and obtaining fetal sexing so that they can have the option of terminating pregnancies of the undesired (usually female) sex. This practice is likely to increase in some countries with the advent of first trimester prenatal diagnosis and it would be helpful to have a consensus view from the non-medical community about the ethics, particularly as it seems to be provided exclusively in the private sector.

The importance of prenatal diagnosis should not be thought of as entirely in the context of selective termination. The earliest disorders to be recognised were the easiest to recognise because they caused the most severe abnormality. Anencephaly, hydrocephaly, severe spina bifida, chromosomal trisomy and severe metabolic defects tend to be untreatable, so that the only benefit that early prenatal diagnosis of these conditions can offer to the pregnant woman and her husband is the option of termination. With the development of more sophisticated techniques, prenatal diagnosis can now be extended to disorders which are potentially treatable *in utero*. Examples include intrauterine bladder catheterisation for urethral obstruction, and maternal digitalisation for certain fetal arrhythmias due to cardiac malformations. In a number of cases, prenatal diagnosis of non-lethal malformations (such as exomphalos) is helpful so that arrangements can be made for the baby to be delivered at a centre where facilities exist for optimal postnatal management.

THE TECHNIQUES

The earliest technique used for the prenatal diagnosis of genetic disorders was sampling the amniotic fluid at about 16 weeks gestation by transabdominal amniocentesis. The first application was to diagnose male fetuses at risk for haemophilia and muscular dystrophy by means of sex chromatin studies on fixed amniotic fluid cells.[17] When it was shown in 1966 that these cells could be cultured,[19] the way was open for the diagnosis of fetal chromosome abnormalities,[10] and inborn errors of metabolism.[16] The first termination of a fetal abnormality following ultrasound diagnosis occurred in 1972,[4] and in the same year Brock and Sutcliffe[3] introduced the measurement of amniotic alphafetoprotein (AFP) for the diagnosis of open neural tube defects. In 1973 fetoscopy was first used to visualise fetal abnormalities,[18] and several groups introduced maternal serum AFP assay as a screening test for pregnancies at risk of open neural tube defects.[2,13] Fetal blood sampling, principally for the diagnosis of haemoglobinopathies, was introduced in 1974 by placentacentesis,[11] or by fetoscopy.[9] Attempts to develop first trimester diagnosis by chorion villus sampling (CVS) were made since 1968,[8] but the successful diagnosis of genetic disorders in continuing pregnancies was not reported until 1982.[12] Both transcervical and transabdominal approaches for CVS are now used.

The advantages and disadvantages of the various techniques currently used for prenatal diagnosis are largely self-evident and depend to a great extent on the indication. The least hazardous technique is *obstetric ultrasound* because it is non-invasive. It is valuable for the visualisation of congenital malformations and for the measurement of organs and limbs in a variety of dysmorphic syndromes including skeletal dysplasias. *Amniocentesis* at 16–18 weeks is essential for the diagnosis of disorders such as open neural tube defects which depend on the analysis of amniotic fluid, and is widely used for chromosomal disorders and metabolic defects. In experienced hands, the risk of fetal loss due to the procedure is in the order of 0.3%, which makes it the least hazardous of the invasive methods. However, termination of pregnancy at 16–20 weeks by prostaglandin induction is often a traumatic experience for the mother, so that an earlier diagnosis would be preferable. *Fetoscopy* carries a fetal loss risk of approximately 4%, but may be the only method available for the diagnosis of certain haematological disorders and for fetal liver and skin biopsy. Fetoscopy has also been used to facilitate fetal exchange transfusion, and to achieve selective feto-cide when only 1 of a pair of dizygotic twins is found to be affected.

CVS appears to have a fetal loss rate of between 2 and 4%, depending on the experience of the operator. It can be used for fetal chromosome analysis and enzyme assay, and is probably the method of choice for all conditions diagnosed by DNA analysis. The great advantage of CVS is that it can be used at betwen 8 and 11 weeks of pregnancy, when direct chromosome analysis can provide a diagnosis within 24 hours. If termination is needed, this can be performed by simple evacuation of the uterus at a stage when only the couple need know of the existence of the pregnancy. It is infinitely more acceptable to the mother than second trimester termination, despite the increased risk of fetal loss associated with CVS. In most cases the mother can start another pregnancy within a few weeks of the termination.

The main disadvantage of CVS at present in Britain is that most mothers who might wish to avail themselves of it are already too far advanced in pregnancy by the time they see their obstetricians. Efforts are required to ensure that women suitable for the procedure are identified and counselled before pregnancy.

THE INDICATIONS, UPTAKE AND IMPACT OF PRENATAL DIAGNOSIS

Fetal Chromosome Aberrations

The most frequent indication for prenatal diagnosis is the risk of fetal chromo-some abnormality in older mothers. For Down's syndrome the average rate among all live-born infants is 0.14%. For mothers aged 35, the risk is 2–3 times this rate; by the maternal age of 40, the risk is 1% and by 43 the risk has doubled again to 2%. In 46 year old mothers the risk is estimated to be about 5%. Rates for the other autosomal trisomies and for sex chromosome abnormalities also increase with maternal age. Accurate livebirth rates are not available, but the European Collaborative Study has published detailed data on the maternal age specific rates for all major chromosome aberrations at the time of amnio-centesis.[7] Rates for all abnormalities increase from 0.9% at age 35 to 10% at age 47, with mothers aged 39 years showing a rate of 2%. It is established that for Down's syndrome these rates are 30% higher than might be expected at birth, as many chromosomally abnormal conceptions are lost by late abortion or stillbirth after 16 weeks gestation.

The knowledge that chromosome aberrations occur more commonly in the offspring of older mothers has prompted many mothers to seek screening by amniocentesis and, more recently, by CVS. As laboratory facilities and other resources are limited, the recommended maternal age for such screening is usually 35 years and over. Some centres with limited resources have been constrained to use a higher cut-off of 37 or 38 years. In most populations the proportion of pregnant women 35 years or over varies from 4 to 7% and so the theoretical maximum reduction in Down's syndrome births which could be achieved, if all women in this age category had prenatal diagnosis and selective termination, would be between 30 and 35%. Only a small number of centres have published on the utilisation and impact of prenatal screening for Down's syndrome and the results show wide variation in practice. In the West of Scotland in 1983, only 32% of women 35 years of age and over were screened and this reduced the total number of births of Down's syndrome by only 14.5% (Table 1). In Merseyside, the only other British centre to publish data, the uptake was 44.1% in 1983 and the percentage of Down's syndrome births avoided was 23%.[20] The best utilisation figures come from Denmark, where the proportion of older pregnant women is much lower than in Britain, and where up to 72% of all pregnant women aged 35 years and over are tested.[14]

There is little information about the reasons for the low uptake of prenatal screening for chromosome abnormalities in Britain. In Glasgow, about 7% of pregnant women decline testing for religious and personal reasons. A further 15% of women are seen too late in pregnancy for amniocentesis and it seems that most of the remaining 46% are not told about the availability of prenatal diagnosis.

From these observations, it is clear that the most effective way of increasing the impact of prenatal screening on the birth incidence of Down's syndrome would be to increase the proportion of older mothers screened. Other approaches include the identification of those younger mothers at increased risk. Two methods have been suggested: First, the use of ultrasound screening to identify subtle changes which are associated with chromosomal syndromes, including intrauterine growth retardation, delayed fetal movements, disproportionate head/trunk ratios and evidence of congenital malformations (such as duodenal atresia). The second suggested method stems from observations that maternal serum AFP levels are significantly depressed in Down's syndrome pregnancies.[5] Thus 21% of pregnancies with Down's syndrome have a maternal serum AFP value below 0.5 MOM (multiples of the medium), compared with approximately 5% of unaffected pregnancies. Strategies can be designed to incorporate the maternal serum AFP value into risks based on maternal age, so that amniocentesis is offered to women younger than 35 only if their combined risk is equal to or more than the maternal age specific risk at 35 years. In this way the number of amniocenteses can be kept at a reasonable level, e.g., 20% of Down's syndrome in mothers under 35 years might be detected at a cost of screening 4% of such mothers.

The great disadvantage of these methods is that they cannot be used in association with first trimester prenatal diagnosis by CVS, as both ultrasonography and maternal serum testing are done in the second trimester.

TABLE 1
West of scotland prenatal diagnosis programme: impact on down's syndrome

	1976	1977	1978	1979	1980	1981	1982	1983	Total
Total births	35225	33984	35018	37714	37651	38201	35784	34714	288291
Number at maternal age ≥ 35 years	2231	2095	1963	2285	2175	2262	2213	2191	17415
Percent	6.3	6.2	5.6	6.1	5.8	5.9	6.2	6.3	6.0
Number screened	185	391	440	478	546	642	674	702	4058
Percent ≥ 35 years	7.6	18.7	22.4	20.9	25.1	28.4	30.5	32.0	23.3
Trisomy 21 terminations	3	4	1	8	4	3	12	11	46
Trisomy 21 live births	45	39	42	41	33	37	21	47	305
Terminations minus third	2	3	1	6	3	2	8	8	33
Total trisomy 21	47	42	43	47	36	39	29	55	338
Rate per 1,000	1.3	1.2	1.2	1.2	1.0	1.0	0.8	1.6	1.2
Percent trisomy 21 avoided	4.3	7.1	2.3	12.8	8.3	5.1	27.6	14.5	9.8

(Updated from Ferguson-Smith[6])

Fetal Neural Tube Defects

The West of Scotland Screening Programme for the prenatal detection of neural tube defects (NTD) in 1983 reached 79% of the 35,000 pregnant women in the area (Table 2). The programme resulted in the termination of 66% of all known NTD pregnancies in this area, and has resulted in a notable reduction in the number of spina bifida admissions for surgery.

Not all regions in Britain manage to screen a large proportion of pregnancies and recent estimates suggest that in England and Wales, in 1983, the overall utilisation rate was about 56%. (Cuckle, personal communication) Most of the liveborn cases missed by the screening programmes are ones where the defect is 'closed' (i.e. covered by skin which prevents diffusion of AFP from the fetal circulation into the amniotic fluid), but these tend to be less handicapped than the 'open' cases.

While it is clear that the marked reduction in birth prevalence of spina bifida is due to prenatal diagnosis, the failure to notify all terminations has led to the assumption that a major contribution to the falling rate has been made by a substantial reduction in the occurrence of this malformation, possibly due to environmental factors. There has also been a tendency to rely more on ultrasound scanning alone as a screening test, so that numbers taking part in AFP screening have tended to drop in the past 2 years. This is unfortunate as there has been no study reported which indicates routine ultrasound is as sensitive a screening test as serum AFP assay.

Other Fetal Abnormalities

There are a number of other indications for prenatal diagnosis, but these are numerically less important than fetal chromosome aberrations and neural tube defects. Over 70 inborn errors of metabolism can be diagnosed from amniotic cell cultures or chorionic villi. The indication is invariably a previous affected sibling and the risk to the fetus is 25%. With the exception of the haemoglobinopathies, less than 200 such pregnancies are at risk in Britain each year.

The prenatal diagnosis of cystic fibrosis, which is the most common autosomal recessive disorder in Western Europe, has an unfortunate history of recurrent disappointments. In the absence of understanding about the nature of the biochemical defect, several diagnostic methods based on secondary effects of the mutation have been tried. The most promising of these depends on the demonstration of reduced levels of intestinal alkaline phospatase in the amniotic fluid.[1,15] It appears that, in the fetus, the condition is associated with partial obstruction of the small bowel (a form of meconium ileus) which reduces the intestinal secretion into the amniotic fluid. This is most frequently evident at about 18 weeks gestation, when an echogenic area in the fetal ileum can often be identified by ultrasound.

Other intestinal enzymes, including Q γ-glutamyl transpeptidase, may also be reduced in the amniotic fluid. The false negative and false positive rate of these

10

TABLE 2

West of Scotland MSAFP programme: proportion of NTD – affected births avoided

Expected year of confinement	1976	1977	1978	1979	1980	1981	1982	1983	1984
Screening period	1–4	5–8	9–12	13–16	17–20	21–24	25–28	29–32	33–36
Total pregnancies	28549†	27782†	35081	37714	37651	38201	35784	34714	35070
Pregnancies screened	6122	11585	17220	22929	26320	27889	28406	27510	26423
Proportion screened (%)	21.4	41.7	49.1	60.8	69.9	73.0	79.4	79.2	75.3
NTD terminations	29	43	63	87	75	107	95	73	70
*All other NTDs	109	109	119	106	80	78	58	38	36
Total NTDs	138	152	182	193	155	185	153	111	106
NTD rate per thousand	4.8	5.5	5.2	5.1	4.1	4.8	4.3	3.2	3.0
Proportion terminated (%)	21.0	28.3	34.6	45.1	48.4	57.8	62.1	65.8	66.0

tests are each in the order of 5%, and so prenatal diagnosis is only useful when there is a high prior risk that the fetus will be affected. In effect, this means that the test should only be used when there has been a previous affected child and the prior risk is 1 in 4. In an increasing number of cases this form of prenatal diagnosis is being replaced by DNA analysis of chorionic villi.

The prenatal diagnosis of Beta-thalassaemia and the other haemo-globinopathies in recent years has been a major success story for our Mediter-ranean colleagues in Northern Italy, Sardinia, Cyprus, and Greece. A major reduction in the birth incidence has been achieved in each of these countries, first through the techniques of fetal blood sampling and globin chain synthesis, and latterly by the application of DNA analysis to CVS.

As has been mentioned, the first prenatal diagnostic test to be used for genetic disorders was fetal sexing for X-linked disease. Nowadays, most of these dis-orders, for example, haemophilia and Hunter's syndrome, have specific prenatal tests to distinguish the affected from the unaffected male fetus. In other con-ditions, like Duchenne muscular dystrophy, linkage studies using DNA polymor-phisms can often make the same distinction with high probability, and so there is less need for terminations on the grounds of fetal sex alone.

Where parental chromosome translocations or inversions are present, the average risk of an unbalanced fetal karyotype is usually between 10 and 15%, and so prenatal diagnosis is imperative. The most suitable technique for these high risk indications is undoubtedly CVS and most of our cases now choose this option provided the carrier state is recognised in good time.

PRENATAL DIAGNOSIS BY DNA ANALYSIS OF CHORIONIC VILLI

The development in medical genetics which holds most promise for the future in prenatal diagnosis is undoubtedly the application of recombinant DNA tech-niques. So far DNA analysis of chorionic villi has had most impact in the diagnosis of Beta-thalassaemia and other haemoglobinopathies, including sickle cell disease, but it has great potential for all single gene defects provided suitable DNA probes can be found. Specific gene probes, consisting of DNA sequences either within or closely adjacent to the gene locus, are now available for the alpha- and beta-globin genes, haemophilia A and B, phenylketonuria and many others. Such probes can be used to identify the specific mutation as in sickle cell disease, or to identify a DNA polymorphism within the gene which can be used as a marker to follow the mutant gene in the family. DNA markers are so common that a study of most families usually reveals a suitable polymorphism which can be used for prenatal diagnosis, always provided that the key relatives are available for testing.

Even when the gene in question has not been cloned, so that there is no specific DNA probe, it is often possible to track a mutant gene in a family using genetic linkage to a DNA marker recognised by a gene probe which maps to the same chromosome region. In this case the gene probe is usually a DNA sequence of unknown ('anonymous') function which has been cloned and subsequently found to show linkage with the disease locus. The most useful probes are those which show the least recombination between the disease and marker loci. Such probes have been found for a number of important diseases, including X-linked muscular dystrophy, Huntington's chorea, myotonic dystrophy, cystic fibrosis and adult polycystic disease of the kidney. A number of centres are already using such probes for the prenatal diagnosis of muscular dystrophy and cystic fibrosis. In many cases, the chance of false-positive and negative results due to meiotic recombination can be reduced by using 2 or more probes which flank the disease locus.

These studies are at an early stage but one can gauge the relative impact that DNA analysis may have on reducing the birth frequency of various single gene defects from what is known about the prevalence of these disorders (Table 3).

TABLE 3

Possible impact of DNA analysis on the prenatal diagnosis of single gene defects

Condition	Regional 35,000 births		National 600,000 births	
	No. at risk	No. affected	No. at risk	No. affected
Cystic fibrosis:				
without carrier detection	16	4	270	67
with carrier detection	88	22	1500	375
Haemoglobinopathies	24	6	400	100
Haemophilia A and B	16	4	240	60
X-L in ked muscular dystrophy	20	5	480	120
Huntington's chorea	30	15	600	300
Adult polycystic kidney disease	40	20	800	400

Pregnancies at risk of cystic fibrosis and other autosomal recessive disorders like phenylketonuria and alpha-1-antitrypsin deficiency are at present only recognised because of a previous affected child; less than 18% of all affected births are in this category, and the impact of prenatal diagnosis on prevention is therefore rather small. Less than 300 pregnancies would be recognised as being at risk for cystic fibrosis in Britain each year. Once it is possible to recognise the 1 in 20 of the population who are carriers and thus the 1 in 400 of all pregnancies which have a 25% risk of cystic fibrosis, the demand for prenatal diagnosis would increase five-fold. Each year 1,500 pregnancies in Britain would require testing to identify the 375 affected with cystic fibrosis.

Carrier detection is already possible for X-linked disorders detectable by DNA probes. However, 30% of those affected with Duchenne muscular dystrophy arise by new mutation and are therefore not avoidable by prenatal diagnosis. The total number of pregnancies identifiable as being at risk is therefore comparatively small, and this applies also to the haemophilias because they are rare disorders. The greatest impact of DNA analysis must be expected in the autosomal dominant group currently exemplified by Huntington's chorea and adult polycystic disease of the kidney. Assuming that families are informative and will agree to predictive testing, it should be possible to identify a high proportion of those who carry the gene in time to provide prenatal diagnosis in each of their pregnancies. For adult polycystic disease of the kidney only rough estimates are possible, but prenatal testing would be required in as many as 800 pregnancies in Britain each year, of which half can be expected to be affected. It remains to be seen how such families will respond to the availability of prenatal diagnosis by CVS and these theoretical estimates are likely to be considerably altered in the light of experience.

CONCLUSION

Prenatal diagnosis is an important adjunct to genetic counselling and both have made an important contribution to the reduction in the burden of genetic disease and severe handicap in the community. Experience with CVS in the first trimester suggests that it may replace second trimester amniocentesis as a more acceptable method for fetal chromosome analysis, biochemical assays and DNA studies. Before this happens, concern about the hazards of the technique must be resolved, and, in many countries including Britain, better strategies are needed for offering this option to women who would benefit from early prenatal diagnosis.

The great potential of DNA methods for the prenatal diagnosis of single gene defects by CVS is illustrated by recent work with X-linked muscular dystrophy, Huntington's chorea, adult polycystic disease of the kidney and cystic fibrosis. The greatest impact of DNA analysis is likely for the autosomal dominant disorders and for those autosomal recessive conditions, like Beta-thalassaemia, in which carrier detection is possible.

The widespread utilisation of prenatal spina bifida screening programmes and the rapid development of new and better diagnostic techniques including DNA analysis, ultrasound visualisation of the fetus and direct chromosome analysis of chorionic villi, attest to the continuing demand and general acceptability of prenatal diagnosis by both the public and the medical profession. Improvements in practice and delivery of services which lead to earlier safe prenatal testing and more rapid provision of results would be particularly welcomed by the families concerned.

BIBLIOGRAPHY

1 Brock, D. J. H., Bedgood, D., Barron, L. and Hayward, C.: 1985, *Prospective prenatal diagnosis of cystic fibrosis,* Lancet, *1*, 1175.
2 Brock, D. J. H., Bolton, A. E. and Monaghan, J. M.: 1973, *Prenatal diagnosis of anencephaly through maternal serum alphafetoprotein measurement,* Lancet, *2*, 923.

14

3 Brock, D. J. H. and Sutcliffe, R. G.: 1972, *Alphafetoprotein in the antenatal diagnosis of anencephaly and spina bifida*, Lancet, *2*, 197.

4 Campbell, S., Pryse-Davies, J., Coltart, T. M., Sellar, M. J. and Singer, J. D.: 1975, *Ultrasound in the diagnosis of spina bifida*, Lancet, *1*, 1065.

5 Cuckle, H. S., Wald, N. J. and Lindenbaum, R. H.: 1984, *Maternal serum alphafetoprotein measurement: a screening test for Down syndrome*, Lancet, *1*, 926.

6 Ferguson-Smith, M. A.: 1983, *Prenatal chromosome analysis and its impact on the birth incidence of chromosome disorders*, Brit. Med. Bull., *39*, 355.

7 Ferguson-Smith, M. A. and Yates, J. R. W.: 1984, *Maternal age specific rates for chromosome aberrations and factors influencing them*, Report of a Collaborative European Study on 52965 amniocenteses. Prenat. Diag., Special Issue, *4*, 5.

8 Hahnemann, N. and Mohr, J.: 1969, *Antenatal fetal diagnosis in genetic disease*, Bull. Europ. Soc. Hum. Genet., *3*, 47.

9 Hobbins, J. C. and Mahoney, M. J.: 1974, *In utero diagnosis of haemoglobinopathies. Technique for obtaining fetal blood*, New Engl. J. Med., *290*, 1065.

10 Jacobson, C. B. and Barter, R. H.: 1967, *Intrauterine diagnosis and management of genetic defects*, Amer. J. Obstet. Gynec., *99*, 796.

11 Kan, Y. W., Valenti, C., Carnazza, V., Guidotti, R. and Fieder, R. F.: 1974, *Fetal blood sampling in utero*, Lancet *1*, 79.

12 Kazy, Z., Rozovsky, I. S. and Bakharev, V. A.: 1982, *Chorionic biopsy in early pregnancy: a method of early prenatal diagnosis for inherited disorders*, Prenat. Diag., *2*, 39.

13 Leek, A. E., Ruoss, C. F., Kitau, M. J. and Chard, T.: 1973, *Raised alphafetoprotein in maternal serum with anencephalic pregnancy*, Lancet, *2*, 385.

14 Mikkelsen, M., Fischer, G., Hansen, J., Pilgaard, B. and Nielsen, J.: 1983, *The impact of legal termination of pregnancy and of prenatal diagnosis on the birth prevalence of Down syndrome in Denmark*, Ann. Hum. Genet., *47*, 123.

15 Muller, F., Berg, S., Frot, J. C., Boué, J. and Boué, A.: 1985, *Prenatal diagnosis of cystic fibrosis. I. Prospective study of 51 pregnancies*, Prenat. Diag., *5*, 97.

16 Nadler, H. L.: 1968, *Antenatal detection of hereditary disorders*, Pediatrics, *42*, 912.

17 Riis, P. and Fuchs, F.: 1960, *Antenatal determination of fetal sex in prevention of hereditary diseases*, Lancet, *2*, 180.

18 Scrimgeour, J. B.: 1973, *Other techniques for antenatal diagnosis*, In, A. E. H. Emery, ed., Antenatal Diagnosis of Genetic Disease, Churchill Livingston, Edinburgh, 40–57.

19 Steele, M. W. and Breg, W. R.: 1966, *Chromosome analysis of human amniotic fluid cells*, Lancet, *1*, 383.

20 Walker, S. and Howard, P. J.: 1986, *Cytogenetic prenatal diagnosis and its relative effectiveness in the Mersey region and North Wales*, Prenat. Diag., *6*, 13.

FIRST TRIMESTER FETAL DIAGNOSIS OF GENETIC DISEASES: TECHNIQUES, RISKS AND INDICATIONS

INTRODUCTION

Hereditary diseases and congenital defects are major causes of fetal and neonatal mortality, later morbidity, and chronic disability including mental retardation. Nevertheless, primary and secondary preventive measures are often possible (Table 1).

TABLE 1

Primary and secondary preventive measures in genetic diseases

---

Primary Measures (before conception)
  A) Preventing mutation
  B) Carrier detection
  C) Genetic counselling, resulting in
    (1) refraining from child-bearing, or
    (2) choosing adoption, or, if husband's genotype is the problem, artificial insemination by donor sperm

Secondary Measures (after conception)
  A) Unselective voluntary abortion
  B) Selective voluntary abortion following fetal diagnosis (if abortion is an acceptable alternative)
  C) Fetal or neonatal diagnosis and institution of appropriate treatment to prevent disability
  D) Gene therapy

Diagnosing an abnormal fetus has become an accepted procedure for at-risk couples, and various methodologies facilitate fetal diagnosis of genetic disease. Removal of a small volume of amniotic fluid is achieved by transabdominal amniocentesis; chromosomal aberrations and some inborn errors of metabolism can then be investigated by culturing the amniocytes for karyotyping and enzyme analysis, respectively. Some disorders can only be diagnosed by obtaining fetal blood (haemoglobinopathies, haemophilia) or skin biopsies (epidermolysis bullosa, ichthyosis) under direct fetoscopic or ultrasonic guidance. Moreover, major structural fetal abnormalities can be recognised by ultrasound, and neural tube defects (NTD) screened by evaluating alpha-fetoprotein (AFP) in maternal blood.

Until recently, the most important limitation of fetal diagnosis was that it involved mid-trimester abortion, usually near 20 weeks' gestation or later. The associated risk of both obstetrical complications[8] and long term maternal psychological sequelae[11] appears to be considerable.

15

E.K. Hicks and J.M. Berg (eds.), The Genetics of Mental Retardation, 15–22
© *1988 by Kluwer Academic Publishers.*

Since 1983 first trimester diagnosis by chorionic villus sampling (CVS) has been available.[6,10,14,15]

Trophoblast accurately reflects the genetic constitution of the fetus and can easily be obtained between 8 and 12 weeks of gestation by the transcervical or transabdominal route.

At present, 4 primary sampling approaches are in use. In 3 of them a catheter[5], biopsy forceps[9], or endoscope[12] is inserted transcervically; in the fourth, transabdominal puncture is used for aspiration.[16] Although comparative data on these sampling procedures are not yet adequate, most testing has to date been accomplished using transcervical catheter aspiration. Over 70% of worldwide sampling has involved the use of a 21 or 26 cm. long and 1.45 mm diameter catheter (Portex Ltd., England),[13] and we have found this system to be the most efficient and reproduceable.[4] Only 3 centres (Thomas Jefferson University, Philadelphia; Michael Reese Medical Center, Chicago; L. Mangiagalli Clinic, Milan) have, to date, sampled about 3,000 cases using this method and following similar criteria.[13] The efficacy of the method, as judged by the total amount of tissue aspirated, and the number of sampling failures, is very similar in different large series (Table 2).

TABLE 2

Efficacy of CVS by transcervical aspiration with a plastic catheter (Portex Ltd, England) in two large series

Centre	Success rate at 1st insertion	Overall success rate	Mean amount of chorion tissue sample per insertion
L. Mangiagalli Clinic, University of Milan	91%	98.9%	35 mg
Thomas Jefferson University, Philadelphia	84,5%	99.5%	25 mg

Conclusive evidence of the risks of CVS must await the collection of more extensive data from collaborative studies. The clinical data from a large centre may, however, provide reliable information about the risk of pregnancy complications.

METHODOLOGICAL PROBLEMS IN CVS

Since CVS is still in an applied research stage, medical centres providing CVS have a responsibility to obtain informed consent. The patient should receive oral and written information on the risk figures for complications, derived from local and international experience, for both first and second trimester procedures.

CVS is an outpatient procedure which takes 15 to 20 minutes.[3] It does not require pre-operative preparation or post-operative care. Post-CVS bleeding has not, to date, had adverse effects, and only bed rest is recommended. The major procedural steps in transcervical CVS (Figure 1) have been standardised, and can be summarised as follows:
(a) ultrasonic evaluation of the uterus and its contents,
(b) cleansing of the external genitalia, vagina and cervix by a broad spectrum antimicrobial solution,
(c) exploration of the cervical canal and identification of the internal cervical os by using a hysterometer,
(d) appropriate bending of the catheter following evaluation of the angle between the cervical and uterine axis, in order to reach the placenta,
(e) gentle insertion of the catheter under continuous ultrasonic visualisation until it has entered the placenta at considerable depth,
(f) after removing the obturator, aspiration of chorionic tissue by a 5–10 ml. depression in a 20–30 ml. syringe containing 2–3 ml. of Hanks' Medium, and
(g) control of the aspirated material under an inverted microscope.

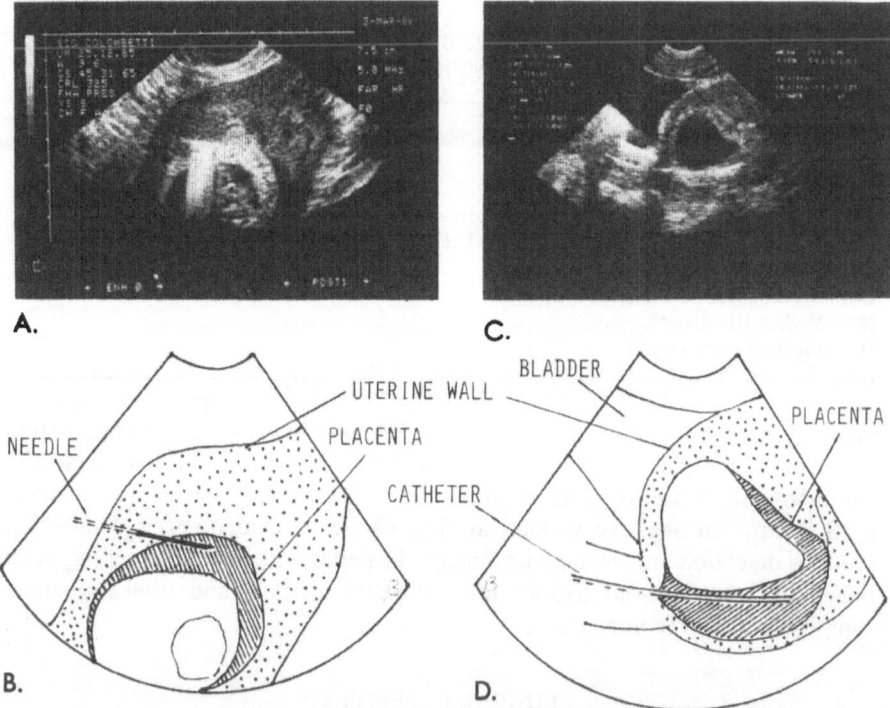

(Top) Ultrasonic sector scan showing transabdominal (A) and transcervical (C) CVS approaches in 2 pregnancies at 10 weeks of gestation. A 20 gauge spinal needle and a 1.5 mm. outer diameter polyethylene catheter were used for the transabdominal and transcervical sampling, respectively.

(Bottom) Explanatory diagrams of the ultrasonic scan (A) and ultrasonic scan (C).

High resolution ultrasound equipment having a sector scanning head is essential, first to confirm fetal viability and allow an accurate evaluation of the gestational sac characteristics, and secondly to guarantee safe and successful sampling through monitoring of the catheter tip. Moreover, there should be sonographic follow-up to monitor for possible CVS sequelae. In addition, AFP measurement in maternal serum can be undertaken between 16 and 20 weeks to obtain information on the risk of NTDs and some other major abnormalities.

The optimal period for obtaining transcervical placental biopsies is between the 9th and the 11th week. Prior to the 9th week, the placenta often cannot be differentiated from the strong echogenicity of the decidual tissue surrounding the gestational sac. Nevertheless, CVS could be performed earlier if the placenta is clearly visualised and its location confirmed by detection of the umbilical cord insertion. As pregnancy advances, the gestational sac increases in size, and completely fills the uterine cavity by 11 weeks. Thereafter, reaching the placenta without damaging membranes is often impossible.

Additional contra-indications of transcervical CVS are summarised in Table 3. In most of these circumstances (A-F), transabdominal aspiration (Figure 1) could be considered as a safer alternative. Because feto-maternal

TABLE 3

Contraindications to transcervical chorionic tissue biopsy Absolute (+ +) and Relative (**)

A)	Clinical Vaginitis + +
B)	Vaginismus + +
C)	Cervical Canal Inaccessibility + +
D)	Myoma Obstructing the Sampling Route + +
E)	Pronounced Angulation of the Corpus Uteri on the Cervix + +
F)	Pregnancy > 11 weeks **
G)	Maternal Rh Isoimmunisation + +
H)	Vaginal Bleeding **
I)	Multiple Pregnancy **

bleeding frequently follows CVS[2], and its effect on the maternal immune-system is yet unknown, an Rh-isoimmunisation should be considered a contra-indication to CVS. Moreover, anti-D immunoglobulins should be administered after CVS to any Rh negative woman at risk. Clinically evident vaginal infection requires detection and specific treatment. Together with vaginal bleeding, such infections are contra-indications for transcervical CVS, and transabdominal sampling can be recommended.

CLINICAL EXPERIENCE

By November, 1985, 910 patients were seen for first trimester fetal diagnosis at the L. Mangiagalli Clinic in Milan. Maternal age was 35 or more in 75% of the cases, with the risk of a fetal chromosomal aberration being the most frequent indication for testing (Table 4). The introduction of DNA technologies gives new impetus to first trimester diagnosis for couples at risk for fetal haemoglobino-pathies[14] or haemophilia.[1] Until recently, only a diagnosis from fetal blood late in the second trimester could detect these disorders. In 3 cases, fetal sex determination was the primary CVS indication, as a result of psychiatric referrals. This small number seems to indicate minimal potential misuse of CVS for non-genetic reasons. Nevertheless, a 'wrong' fetal sex could be the motivation for terminating a pregnancy shortly after receipt of normal test results in the majority of non-genetically based abortion cases (Table 5). Contradictory conscious or unconscious motivations are often present in the early stages of gestation, and frequently play an important role in the decision to continue or terminate a pregnancy. This suggests the necessity of having a psychotherapist permanently on the counselling team.

TABLE 4

Indications for CVS at L. Mangiagalli Clinic, University of Milan

Chromosomal Abnormalities		817
Metabolic Diseases		16
Fetal Sex Determination:		
X-linked diseases	48	
Other reasons	7	55
Haemoglobinopathies		20
Haemophilia A		2
TOTAL		910

TABLE 5

Clinical experience with CVS at L. Mangiagalli Clinic, University of Milan

Total number of patients	910	
Twin pregnancies	6	
Voluntary abortion:	62	
genetic 49 (5.4%)		
non-genetic 13 (1.4%)		
Pregnancies intended to continue	848	
Completed pregnancies	615	
Fetal loss:		
≤20 weeks 20 (3.2%)	20	(3.2%)
>20 weeks <28 weeks 5 (0.8%)	5	(0.8%)
*Stillbirth	3	
*Neonatal death (≤7 days)	4	

* perinatal mortality (1.2%)

CVS was performed at 9 or 10 weeks in 85% of the cases, and a successful sampling was obtained in over 90% at first catheter insertion, with an overall success rate of 98.9% at the second insertion (Table 2). In 25 of 615 completed pregnancies (Table 5) fetal loss occurred prior to 28 weeks (4.0%), and the overall loss rate, including stillbirth and neonatal death, was 5.2% (accounting for the perinatal mortality of 1.2%). Fetal loss is much lower after exclusion of the 7 cases of loss which were unrelated to CVS (Table 6).

TABLE 6
Fetal losses unrelated to CVS

Case number	Gestational age at abortion or delivery	Weeks post-CVS	Clinical comments
54	10	1	Missed Abortion; fetal karyotype 47,XX, + 14
144	9	1	Missed Abortion; fetal karyotype 45,X
149	39	29	Diaphragmatic Hernia; neonatal death
154	33	22	Chronic Amniotic Fluid leakage following fetoscopy for Haemophilia A diagnosis; spontaneous labour; fetal lung hypoplasia
254	37	26	Meconium aspiration syndrome
275	39	30	Meconium aspiration syndrome
388	21	12	Fetal death after fetal blood sampling to conform fragile X

Statistical analysis was undertaken by comparing fetal loss or early complications with procedural variables, i.e., gestational weeks at sampling, number of catheter insertions, weight of chorionic tissue sampled, result of cervical culture, maternal AFP level, and vaginal bleeding. Only multiple catheter insertions (2 or more) were significantly associated with fetal loss. Bacteria and viruses are strongly suspected to be important causes of pregnancy wastage through spontaneous abortion or intrauterine fetal death. Our failure to demonstrate any such relationship might be due to inadequate investigation in this respect. Actually, 2 cases of acute uterine infection, marked by fever, uterine contractions, and spontaneous evacuation have been reported by ourselves,[7] and some by other investigators.[13]

The remarkable number of completed pregnancies provides interesting data on major late complications (Table 7). Intrauterine growth retardation, premature delivery, premature rupture of membranes followed by preterm delivery, and placental disorders were observed in the expected ranges. Moreover, normal intrauterine fetal growth was observed during the second and third trimester by means of serial ultrasonic measurements of the abdominal circumference in 150 CVS pregnancies. Furthermore, the birth weights (plotted against gestational age) of children born after CVS had median values very similar to those from

TABLE 7

Complications following CVS observed at the L. Mangiagalli Clinic,
University of Milan

EARLY COMPLICATIONS:		
Total number of continuing pregnancies	848	
Vaginal bleeding	104	(12.3%)
Intrauterine haematoma	39	(4.7%)
Perforation of membranes	0	
Intrauterine infection 2	2	(0.2%)
LATE COMPLICATIONS:		
Total number of delivered pregnancies	615	
Premature delivery (<37 weeks)	38	(6.2%)
IUGR (weight <10th percentile)	52	(8.5%)
(weight <5th percentile)	26	(4.2%)
PRM and preterm delivery	5	(0.8%)
Abruptio placentae	6	(1.0%)
Placenta previa	4	(0.7%)
Placenta accreta	1	(0.2%)

IUGR = intrauterine growth retardation
PRM = premature rupture of membrane

the Italian standards, with a Gaussian distribution demonstrable from the 38th to the 41st week.

The incidence of congenital defects found neonatally (2.6%) was not increased, and no uncommon abnormalities were recognised. Major anomalies were present in 2 of 11 cases (1 diaphragmatic hernia, and 1 ascites and liver malformation); in the remaining 9 cases a minor defect was found (club foot in 4 instances, and cutaneous syndactyly, flat angioma, abnormal shape of auricle, auricular agenesis, and short umbilical cord in 1 case each).

In conclusion, our experience indicates that: (1) CVS seems to be a safe and reliable option for first trimester prenatal diagnosis in highly skilled centres which also provide extensive counselling. Although the experience of such centres is promising, conclusive assessment of the accuracy and the risks associated with CVS requires further collaborative studies. (2) Only obstetricians and gynaecologists experienced in this procedure (after obtaining skills in experimental cases) should be permited to do CVS. This would facilitate successsful sampling at the first insertion in approximately 90% of the cases, and permit an overall success rate of nearly 100% at second insertions. Additionally, transabdominal sampling should be available as a complementary approach. (3) Although intrauterine infection is an unusual complication of CVS (occurring at a rate of 2–3 per 1,000), it should be referred to as a possibility in genetic counselling. (4) Ultrasound follow-up of fetal growth, and birth weight distribution, indicate normal patterns in CVS monitored pregnancies. (5) Neither pregnancy complications nor the incidence of congenital defects exceed the expected rate.

22

BIBLIOGRAPHY

1 Antonarakis, S. E., Copeland, K. L., Carpenter, R. J., Carta, C. A., Hoyer, L. W., Caskey, C. T., Toole, J. J. and Kazazian, H. H.: 1985, *Prenatal diagnosis of haemophilia A by factor VIII gene analysis*, Lancet, *1*, 1407.

2 Brambati, B., Guercilena, S., Bonacchi, I., Oldrini, A., Lanzani, A. and Piceni, L.: 1986, *Feto-maternal transfusion after chorionic villus sampling: clinical implications*, Hum. Reprod., *1*, 37.

3 Brambati, B. and Oldrini, A.: 1986, *Methods of chorionic villus sampling*, In, B. Brambati, G. Simoni and S. Fabro, eds., Chorionic Villus Sampling, Marcel Dekker, Inc., New York, 73–98.

4 Brambati, B., Oldrini, A. and Aladerun, S. A.: 1983, *Methods of chorionic villi sampling in the first trimester fetal diagnosis*, In, A. Albertini and P. G. Crosignani, eds., Progress in Prenatal Medicine, Excerpta Medica, Amsterdam, 275–283.

5 Brambati, B., Oldrini, A., Ferrazzi, E. and Lanzani, A.: 1985, *Chorionic villi sampling: general methodological and clinical approach*, In, M. Fraccaro, G. Simoni, and B. Brambati, eds., First Trimester Fetal Diagnosis, Springer Verlag, Berlin, 7–18.

6 Brambati, B. and Simoni, G.: 1983, *Diagnosis of fetal trisomy 21 in first trimester*, Lancet, *1* 586.

7 Brambati, B. and Varotto, F.: 1985, *Infection and chorionic villus sampling*, Lancet, *2*, 609.

8 Cates, W. and Grimes, D. A.: 1978, *Deaths from second trimester abortion by dilatation and evacuation: causes, prevention, facilities*, Obstet. Gynec., *58*, 401.

9 Dumez, Y., Goossens, M., Boué, J., Poenaru, L., Dommergues, M. and Henrion, R.: 1985, *Chorionic villi sampling using rigid forceps under ultrasound control*, In, M. Fraccaro, G. Simoni, and B. Brambati, eds., First Trimester Fetal Diagnosis, Springer Verlag, Berlin, 38–45.

10 Grebner, E. E., Wapner, R. J., Barr, M. A.and Jackson, L. G.: 1983, *Prenatal Tay-Sachs diagnosis by chorionic villi sampling*, Lancet, *2*, 286.

11 Greer, H. S., Lal, S., Lewes, S. C., Belsey, E. M. and Beard, R. W.: 1976, *Psychological consequences of therapeutic abortion*, Kings Termination Study III, Brit. J. Psychiat., *128*, 74.

12 Gustavii, B.: 1983, *First trimester chromosomal analysis of chorionic villi obtained by direct vision technique*, Lancet, *2*, 507.

13 Jackson, L. G.: 1985, *CVS Newsletter*, Philadephia, December 9.

14 Old, J. M., Ward, R. T. H., Karagozlu, F., Petrou, M., Modell, B. and Weatherall, D. J.: 1982, *First trimester fetal diagnosis for haemoglobinopathies: three cases*, Lancet, *2*, 1414.

15 Pergament, E., Ginsburg, N., Verlinsky, Y., Cadkin, A., Chu, A. and Trnka, L.: 1983, *Prenatal Tay-Sachs diagnosis by chorionic villus sampling*, Lancet, *2*, 286.

16 Smidt-Jensen, S. and Hahnemann, N.: 1984, *Transabdominal fine needle biopsy from chorionic villi in the first trimester*, Prenat. Diagn., *4*, 163.

M. MIKKELSEN

CHROMOSOMAL FINDINGS IN FIRST TRIMESTER CHORIONIC VILLI BIOPSY

INTRODUCTION

Second trimester amniocentesis and selective abortion of fetuses with severe genetic disease has been carried out in Western countries for more than 15 years. Especially in societies with a long tradition of legal abortions on genetic grounds, utilisation of such services has been high.[8] Nevertheless, the interruption of a wanted pregnancy relatively late in gestation because of a fetal disease has been a psychological trauma for the women in question. Many parents at risk of bearing offspring with severe disease have suffered profound conflicts.

There is no doubt that fetal diagnosis and the possibility of induced abortion in the first trimester would be less traumatic to the parents. There is less bondage between mother and fetus, the abortion procedure is simpler and safer, and the waiting time before a diagnosis is available is much shorter.

There are, nevertheless, a number of questions to be answered before the first trimester approach can replace amniocentesis. These questions can be answered in a limited time period only through international cooperation. The most important question is the safety of the procedure (dealt with by B. Brambati, this volume). The second question is its reliability, and the third concerns risk figures in the first trimester. This paper deals with the last two questions.

MATERIAL AND METHOD

Data on chorionic villi sampling (CVS) were collected for the Rapallo meeting in October, 1984.[6] Additional information has since reached me. To date, detailed analyses of the cytogenetic findings are available for more than 3,000 chorionic villi biopsies, with more than 30 centres in 13 countries participating in this collaborative study. These data are given in the tables. All comparisons with amniocentesis or newborn data were made by X^2 tests. Several studies published in 1984 were the sources of the amniocentesis[1,2,11], and of the newborn[10], data.

RESULTS

Advanced Maternal Age

The different indications for chromosome investigation and the proportion with abnormal karyotypes are given in Table 1. Eighty percent of the cases were studied because of advanced maternal age, and in nearly 5% a chromosomal aberration was observed. This is significantly higher than in amniocentesis data.

23

E.K. Hicks and J.M. Berg (eds.), The Genetics of Mental Retardation, 23–31
© *1988 by Kluwer Academic Publishers.*

TABLE 1

Chromosome studies in chorionic villi

Indication	No. Studied	%	No. Abnormal	%
Advanced maternal age	2887	81.03	136	4.71
Previous chield/fetus with trisomy 21	260 ⎤		5	1.92
Previous child/fetus with other chromosome aberrations	75 ⎦ 335	9.40	2	2.67
Parental Robertsonian translocation	62 ⎤		8	12.90
Parental reciprocal translocation	52 ⎮ 126	3.54	12	23.08
Other parental aberra-tions	12 ⎦		2	16.67
Parental anxiety and miscellaneous	215	6.03	4	1.86
Total	3563	100.00	169	4.74

As expected, the highest risk figures were found in the group of parents carrying a chromosome aberration themselves. The lowest risk figure was found in the miscellaneous group which also included particularly anxious mothers.

In Table 2 crude data for advanced maternal age are given in single year intervals, with the different types of aberrations found. In the 35–39 year age group, 1,639 fetuses were studied – in 57 cases, a chromosome aberration was observed (3.48%); trisomy 21, including 1 mosaic, was found in 20 cases (1.22%), which is significantly higher than in the amniocentesis and newborn data. In the age group 40 and older, 597 fetuses were studied – 44 abnormal karyotypes were observed (7.37%); trisomy 21 was found in 21 cases (3.52%), also significantly higher than in the amniocentesis and newborn data.

Recurrence Risk For Non-inherited Chromosomal Abnormalities

Recurrence risk figures are given in Table 3. These figures are not significantly different from the amniocentesis data for trisomy 21.

Translocations

Robertsonian translocations involving chromosome 21 showed nearly 20% Down's syndrome fetuses when the mother was the carrier (Table 4). This figure was significantly higher than in the amniocentesis and newborn data. No paternally derived translocation trisomy 21 was observed when the father was the translocation carrier. This finding was exactly the same as in the amnio-

TABLE 2

Advanced maternal age: crude data for single year intervals

Age	No.	Trisomy 21	Trisomy 18	Trisomy 13	Other autosomal trisomies*	Other autosomal aberrations	Sex chromosome aberrations	Totals
35	271	–	–	–	–	1	2	3
36	314	1	–	–	–	–	4	5
37	384	5	–	–	5	–	5	15
38	368	8	2 (1)	–	–	4	3	17(1)
39	302	5 + 1*	2	–	3	3	2	16
40	226	1	–	1	3	–	4	9
41	163	5	–	–	3	–	–	8
42	101	6	1	–	1	2	1	11
43	46	5	1	2	1	–	–	9
44 and more	61	4	1	–	1	1	–	7
Totals	2236	41	7(1)	3	17	11	21	100(1) (4.47%)

* Other automal trisomies; + 14 (1 case); + 15 (6 cases), + 16 (4 cases), + 22 (5 cases), triploidy (1 case)

** Mosaic case

The case in parentheses was a case of trisomy in villi not refound in the fetus. It has been excluded in the calculations of risks

TABLE 3

Recurrence risks: crude data

	No.	Trisomy 21	Trisomy 18	Trisomy 13	Other aberrations	Total	%
Trisomy 21	295	1	1	–	6	8	2.71
Trisomy 18	44	–	1	–	2	3	6.82
Trisomy 13	15	–	–	1	1	2	13.33
Other autosomal aberrations	42	–	–	–	1	1	2.38
Sex chromosome aberrations	22	–	–	–	–	–	–
Totals	418	1	2	1	10	14	3.35

Viable trisomies = 0.96%

TABLE 4

Robertsonian translocations involving chromosome 21

Type of translocation	Mother carrier outcome			Father carrier outcome		
	Normal	Carrier	Unbalanced	Normal	Carrier	Unbalanced
t(13;21)	4	3	0	–	–	–
t(14;21)	10	10	4	3	3	0
t(15;21)	0	0	2	1	1	0
t(21;22)	1	5	2	2	0	0
Total	15	18	8 (19.5%)	6	4	0

TABLE 5

Parental Robertsonian translocations not involving chromosome 21 and parental autosomal reciprocal translations

Type of translocation	Outcome		
	Normal	Carrier	Unbalanced
Robertsonian not involving chromosome 21	4	7	0
Reciprocal autosomal	18	22	12 (23.1%)

No significant sex differences

centesis data. Although unbalanced karyotypes were not observed when chromosome 21 was not involved in the parental Robertsonian translocation (Table 5), the numbers were small. When one of the parents carried a reciprocal translocation, in 23% of the fetuses a chromosome aberration with an unbalanced karyotype was found (Table 5).

X-linked Disorders

Table 6 provides information on X-linked disorders, Duchenne's muscular dystrophy and haemophilia A being the most common ones in which chorionic villi chromosome examinations were undertaken. In 7 of the 306 cases (2.29%), a chromosome aberration was diagnosed, and in 147 cases the sex was male. Not all of the male fetuses were ultimately aborted.

TABLE 6
X-linked disorders: results of chromosomal study of chorionic villi

	46,XX	46,XY	Abnormal karyotype	Totals
Duchenne's muscular dystrophy	63	65	6	134
Becker's muscular dystorphy	2	2	–	4
Haemophilia A	40	40	–	80
Haemophilia B	4	1	–	5
X-linked mental retardation	10	10	–	20
Fragile X syndrome	13	9	–	22
Menkes syndrome	6	6	–	12
Other X-linked disorders	14	14	1	29
Totals	152	147	7 (2.29%)	306

Discordant Findings In Villi And Fetal Cells

As discrepancies between findings in villi and in fetal tissue have been observed[5,9] it was suggested at the Rapallo meeting that efforts be made to study the fetus and placenta when a chromosome aberration is diagnosed in chorionic villi. The results of this collaborative effort are given in Tables 7, 8, and 9. Discrepancies were observed in 2 of the 60 fetuses having an autosomal trisomy (Table 7). One case of trisomy 16 in villi showed a fetal karyotype of 69,XXX. Another fetus, aborted because of trisomy 22 diagnosed in villi, showed a normal karyotype. In the case of an abnormal chromosome 18, it was refound in the fetus, but the placenta showed a normal karyotype. One case of tetraploidy observed in villi was not refound in the amniotic fluid cell culture.

TABLE 7

Autosomal trisomies and other abnormalities observed in chorionic villi, and follow up studies

Type of aberration	No of cases	Confirmed in fetal cells	Not studied (sp.ab, or culture failed)	Placental findings	Discrepancies
Trisomy 21	39	32	6	9	0
Trisomy 18	7	4	0	3	0
Trisomy 13	3	3	0	3	0
Trisomy 14	1	–	sp.ab.	–	–
Trisomy 15	4	–	2	2	0
Trisomy 16	4	1	1	–	1*
Trisomy 22	4	1	1	–	1**
Triploidy	4	1	2 sp.ab.	–	–
Tetraploidy	1	–	–	–	1
Double trisomy	1	–	sp.ab.	–	–
Abnormal chromosome 18	1	1	–	normal	1

* 69,XXX in fetus
** Normal karyotype (i.e. in fetus)
  sp.ab. = spontaneous abortion

TABLE 8

Autosomal mosaics: cytogenetic findings in chorionic villi and outcome of fetal studies

Type of aberration	No. of cases	Confirmed in fetal cells	Confirmed in placenta	Not studied*	Discrepancies No.
46/47 + 3	2	0	1	–	2
46/47 + 5	1	0	–	–	1
46/47 + 7	3	0	–	2 (1 sp.ab.)	1
46/47 + 12	1	0	–	–	1
46/47 + 18	2	0	–	–	2
46/47 + 20	1	0	–	1	1
46/47 + mar	3	1	1	–	2

* failed, spontaneously aborted or not retrieved

Of the 13 autosomal mosaics (Table 8), 3 could not be studied. One case, with a marker chromosome, was confirmed in fetal cells, but in 9 cases the findings in the fetus or infant were normal. Also, sex chromosome mosaics (Table 9) in most cases showed normal chromosomes in the fetus or infant. The XXX, XXY, and 45,X cases were all confirmed in fetal cells or in cord blood after delivery.

TABLE 9

Sex chromosome aberrations: cytogenetic findings in chorionic villi and outcome of fetal studies

Type of aberration	No. of cases	Confirmed in fetal cells	Confirmed in placenta	Not studied	Discrepancies No.
XXX	1	1	–	–	0
XXY	2	2	2	–	0
45,X	4	4	–	–	0
46,XYq +	2	–	–	–	2
45,X/46,XY	2	1	–	–	1
45,X/46,XX	1	–	–	–	1
46,XX/47,XXX	1	–	–	–	1
45,X/46,X/ 56,Xi(Xq)	1	–	–	–	1

DISCUSSION

The data collected for the Rapallo meeting indicated a significantly higher number of chromosome aberrations compared to amniocentesis data at 16 weeks of pregnancy.[6] Additional data collected since then have shown lower prevalences of chromosome abnormalities for the CVS material. Most likely, there are two reasons for this. The first is that CVS was carried out earlier in pregnancy in the initial series. There were a number of cases which had already been studied at the 7th and 8th week of gestation. A doubling of the aberration rate was observed in the 7th-8th week compared to the 10th-11th week. At the Rapallo meeting, it was agreed that sampling should be limited to the period between the 9th and 11th weeks. The second reason is the lower reporting of mosaics in the second part of the study, especially of the types 45,X/46,XY or 45,X/46,XX. Most likely, the quality of the preparations has improved and 45,X lines have been recognised as artefacts. Also, clones with additional chromosomes or with rearrangements have not been reported as aberrations, but have correctly been classified as pseudomosaics.

The large data collection and the collaborative efforts to study all aborted conceptions facilitate the following conclusions. In cases of mosaicism, abortion should not be carried out before the mosaic has been proved in amniotic fluid cell culture. Moreover, rare mosaics with a cell line having a nonviable trisomy could perhaps be disregarded when ultrasonography indicates normal fetal development. This is currently done where trisomy 20 is found at amniocentesis. Trisomy 20, with few exceptions, is regarded as a finding restricted to the chorion. The fetus is apparently less tolerant to chromosomally abnormal cells than the placenta. Also, tetraploidy and tetraploidy mosaics may be artefacts. One case of a tetraploidy mosaic has, however, been rediscovered in the fetus.

There is a difficult decision to make when a mosaic with a viable trisomy is found, e.g. trisomy 21, trisomy 13, or trisomy 18 mosaics. Here, the advantage of first trimester abortion has to be compared to the risk of second trimester amniocentesis with late abortion, and the risk of aborting a normal fetus in early pregnancy. In cases of long lasting infertility or highly advanced maternal age, the choice might be particularly difficult.

It is a serious problem that chromosome abnormalities in villi, whether they are found as mosaics or in the entire sample, may not reflect the karyotype of the fetus. The most likely explanation for the previously mentioned case with trisomy 16 in CVS and triploidy in the fetus, and the case of trisomy 22 in CVS and a normal fetal karyotype, is the concept of a vanishing twin. Additionally, it is unknown what significance a chromosomally abnormal placenta might have on the development of a fetus. Two chromosome aberrations were found in the placentas of small for-date infants studied by Kalousek and Dill.[4] Kalousek[3] outlined the complexity of early embryonic development and indicated that mosaicism confined to the chorion (which has been documented several times) or to the embryo (which was found in the follow-up of the abnormal chromosome 18 case in this study) might exist. This study supports her postulations. Additional data would be advantageous, and fetus and placenta should be studied in all available cases.

The findings of 20% Down's syndrome fetuses in the first trimester when the mother was the carrier of a D or G/21 translocation and zero when it was the father, make it most likely that selection against the unbalanced offspring is postzygotic in the female, while it is prezygotic in the male, carrier.

The finding of 2.29% chromosome aberrations in pregnancies studied because of an X-linked disease shows the importance of a full karyotype in the sex diagnosis instead of using only Y-linked probes.

There is no doubt that first trimester fetal diagnosis has made gynaecologists and geneticists aware of the lack of knowledge of very early human pregnancy. There is especially a lack of baseline data for chromosomal aberrations in fetus and/or placenta in the different maternal age groups in the first 3 months of gestation. Also, the number of twin pregnancies, and how often one of the twins vanishes, is unknown. Such data would also be important for genetic counselling purposes. Many pregnancies end in spontaneous abortions because of a chromosome aberration in the fetus. It would be important to know the natural history from conception to adult life for individuals with chromosomal disorders.

There is also no doubt that first trimester fetal diagnosis is of great advantage to families at high risk of genetic disorders, e.g. X-linked disorders, metabolic diseases and translocation families. Most likely the first trimester approach will also replace amniocentesis in cases of lower risk, e.g. advanced maternal age, or recurrence of a non-inherited chromosome aberration.

An important question is whether the diagnosis early in pregnancy will make both parents and society less tolerant of handicapped persons, even those having potentially minor handicaps. Only a broad information base and discussion can prevent such a development. There has been concern, especially among politicians, that the early approach can be used, or rather misused, in sex selection in connection with family planning. This is possible in societies with freely available abortion, as has been the case in China and in a few instances in Europe. In Denmark, where abortion on demand is legal before the 12th week, it was agreed not to disclose the sex of the fetus prior to the 13th week of pregnancy. This has been criticised as paternalistic, but was necessary for political reasons, to assure that knowledge of the fetal sex was not misused. Fetal diagnosis offers alternative solutions to those parents at risk of having offspring with serious disabilities, and its use should be restricted to this group.

Since the writing of this paper more extensive data have been collected. These data were presented at the 7th International Congress of Human Genetics, held in Berlin in 1986, and appear in the proceedings publication of that congress. [7]

BIBLIOGRAPHY

1 Boué, A. and Gallano, P.: 1984, *A collaborative study of the segregation of inherited chromosome structural rearrangements in 1356 prenatal diagnoses,* In, J. L. Hamerton and M. A. Ferguson-Smith eds., Prenat. Diag., Special Issue, *4*, 45.

2 Ferguson-Smith, M. A. and Yates, J. W. R.: 1984: *Maternal age specific rates for chromosome aberrations and factors influencing them: Report of a collaborative European study on 52965 amniocenteses,* In, J. L. Hamerton and M. A. Ferguson-Smith eds., Prenat. Diag., Special Issue, *4*, 5.

3 Kalousek, D. K.: 1985, *Mosaicism confined to chorionic tissue in human gestations,* In, M. Fraccaro, G. Simoni and B. Brambati, eds., First Trimester Fetal Diagnosis, Springer Verlag, Berlin 130–136.

4 Kalousek, D. K. and Dill, F. J.: 1983, *Chromosomal mosaicism confined to the placenta in human conceptions,* Science, *221,* 665.

5 Karkut, I., Zakrzewski, S. and Sperling K.: 1985, *Mixed karyotypes obtained by chorionic villi analysis: mosaicism and maternal contamination,* In, M. Fraccaro, B. Simoni and B. Brambati, eds., First Trimester Fetal Diagnosis, Springer Verlag, Berlin, 144–146.

6 Mikkelsen, M.: 1985, *Cytogenetic findings in first trimester chorionic villi biopsies: A collaborative study,* In, M. Fraccaro, G. Simoni and B. Brambati, eds., First Trimester Fetal Diagnosis, Springer Verlag, Berlin, 109–120.

7 Mikkelsen, M., and Aymé, S.: 1987, *Chromosomal findings in Chorionic villi: A collaborative study,* In, F. Vogel and K. Sperling, eds., Human Genetics, Proc. 7th Internat. Congr., Berlin, Springer Verlag, Berlin.

8 Mikkelsen, M., Fischer, G., Hansen, J., Pilgaard, B. and Nielsen, J.: 1983, *The impact of legal termination of pregnancy and of prenatal diagnosis on the birth prevalence of Down syndrome in Denmark,* Ann. Hum. Genet., *47,* 123.

9 Simoni, G., Gimelli, G., Cuoco, C., Terzoli, G. L., Rosella, F., Romitti, L., Dalprà, L., Nocera, G., Tibiletti, M. G., Tenti, P. and Fraccaro, M.: 1985, *Discordance between prenatal cytogenetic diagnosis after chorionic villi sampling and chromosomal constitution of the fetus,* In, M. Fraccaro, G. Simoni and B. Brambati, eds., First Trimester Fetal Diagnosis, Springer Verlag, Berlin, 137–143.

10 Stene, J. and Mikkelsen, M.: 1984, *Down syndrome and other chromosome disorders,* In, N. J. Wald, ed., Antenatal and Neonatal Screening, University Press, 74–105.

11 Stene, J., Stene, E. and Mikkelsen, M.: 1984, *Risk for chromosome abnormality at amniocentesis following a child with a non-inherited chromosome aberration,* In, J. L. Hamerton and M. A. Ferguson-Smith, eds., Prenat. Diag., Special Issue, *4*, 81.

M. E. FERGUSON-SMITH

A LABORATORY VIEW OF PRENATAL DIAGNOSIS

Prenatal diagnosis by amniocentesis at 16 weeks gestation has been available since 1967, and by chorionic villi sampling (CVS) in the first trimester of pregnancy since 1982. The availability and extent of the prenatal diagnostic service offered differs greatly around the world and depends largely on the public awareness of genetic disease and severe handicap, the moral attitudes to handicap and abortion and the distribution of public or private funds in prevention of such handicap. In countries where genetic counselling and prenatal diagnosis is part of preventative medicine, particularly in the United States, the legal rights of the mother and the fetus are clearly defined.[7] The volume referred to deals with guidelines concerning the moral and legal obligations of the genetic counsellor or other specialist towards the patient. There are, however, few or no guidelines concerning the basic standards expected from a laboratory where these tests are carried out, or from the laboratory scientists who interpret them. The majority of centres offering prenatal diagnosis evolved either by extending already existing genetic services or from smaller research establishments; therefore, the degree of competence of each laboratory can vary greatly.

Prenatal diagnostic tests in Europe, the United States and Canada, as reported in 1984 in a special issue of Prenatal Diagnosis,[5] are carried out mainly for maternal age and other chromosome-related indications. There are now a considerable number of years of experience in cytogenetic analysis and tissue culture, and more is known about the stress and anxiety felt by the mother between the time of sampling and the time when the results become available. Certain basic standards can now be expected and demanded from a centre before prenatal disgnosis is offered to the community at risk. What should these standards be?
1. Reliable and safe identification of the diagnostic sample
2. Reliability of tissue culture methods
3. Availability of results within a reasonable time
4. Expertise in interpretation of results
5. Back-up facilities for further genetic studies
6. Follow-up facilities and confirmation of diagnosis in genetically affected terminations
7. Confidentiality of records
Each of these points is considered in turn below.

E.K. Hicks and J.M. Berg (eds.), The Genetics of Mental Retardation, 33–37
© *1988 by Kluwer Academic Publishers.*

RELIABLE AND SAFE IDENTIFICATION OF THE DIAGNOSTIC SAMPLE

Most human and clerical errors involving misdiagnosis can occur at the beginning of the testing process when the sample may be wrongly labelled immediately after amniocentesis, or at booking in when the paperwork and sample may be mixed up with another, or at setting up of culture when more that 1 case is handled at a time. Contamination of cultures with another sample either at setting up, feeding or harvesting time is another source of error. If procedures for handling each case are clearly defined and followed, such mistakes can be avoided.

RELIABILITY OF TISSUE CULTURE METHODS

The techniques, materials and equipment for amniotic fluid cell culture have improved greatly in recent years. The United Kingdom External Quality Assessment Scheme[10] reported an overall culture success rate of 97%, with a repeat rate of amniocentesis at about 1.5%. The most common causes for failure to report were the loss of cultures due to contamination or failure of cells to become established in culture.

Prenatal diagnosis by CVS in the first trimester of pregnancy introduced a new dimension to cytogenetic analysis. It allows results to be available within 24 hours if using the direct method pioneered by Simoni and his colleagues in Italy.[8] The direct method is still limited to the extent that some very small or complex chromosomal translocations cannot be accurately diagnosed because of existing limitations in banding techniques. However, short term culture methods as used in Denmark in Dr. M. Mikkelsen's laboratory and in other centres, introduced alongside the direct methods, can provide quick and reliable diagnosis.

If a pregnancy is advanced in gestation and the risk of abnormality is high, fetal blood sampling by fetoscopy can be recommended. The procedure is more hazardous, but the results are available within 72 hours. It is important to have access to all available diagnostic methods to avoid the possibility of a complicated and very distressing termination at an advanced gestation after fetal movements and a fetal-maternal bond have been established.

AVAILABILITY OF RESULTS WITHIN A REASONABLE TIME

Amniocentesis is performed mainly between 16 to 18 weeks gestation, and the results are available within 10 to 21 days. The United Kingdom External Quality Assessment Scheme[10] quoted the average time from receiving to reporting a case as 20.7 days for optimum specimens, and 22.2 days for suboptimum ones. In our centre, as soon as an abnormality is diagnosed, a personal telephone report is made to the clinician. This allows him to contact the patient to give her as much time as possible to adjust to the new circumstances, and if necessary to get further information and counselling regarding that particular abnormality before the final decision is made about termination. Some mothers elect to continue with the pregnancies when abnormalities involving the sex chromosomes are diagnosed. If a case will take longer to report, the clinician should be notified regarding the conditions and reasons for delay. In cases of biochemical defect and late amniocentesis originating from maternal serum screening programmes for neural tube defects, the results are not available until 22 or 23 weeks gestation. With new developments in biochemical and DNA techniques in CVS and better ultrasound facilities, late terminations should decline. All technical and clerical effort must be made to reduce the waiting time, so that termination of an abnormal fetus is a less harrowing and disturbing procedure.

EXPERTISE IN INTERPRETATION OF RESULTS

One of the major problems in interpreting the fetal karyotype in amniotic fluid cultures is chromosomal mosaicism and how it affects the fetal phenotype. Hsu and Perlis,[6] in a United States survey, reported the average frequency of mosaicism involving multiple cells or clones as 0.7% and pseudomosaicism involving single cells or clones as 2.4%. In our department, Ferguson-Smith[3] found pseudomosaicism in 0.45% of cases analysed. If it is impossible to exclude true mosaicism, further cytogenetic studies should be undertaken to confirm the fetal karyotype, either by repeating the amniocentesis or by fetal blood sampling. Ferguson-Smith and Ferguson-Smith[2] reported that in some cases a clear distinction between true and pseudomosaicism cannot be made and this must be appreciated by all clinicians involved in prenatal diagnosis counselling.

Unsuspected chromosome rearrangements and translocations create another problem in the interpretation and reporting of fetal results. Parental bloods must be studied and all available and informative banding techniques utilised. In deciding what effect such an abnormality has on the baby's phenotype, experience from newborn and abortion studies should be employed.

Interpretation of unbanded chromosome results from CVS can be more difficult. For instance, a 47, XX trisomy 21 can be mistaken for a 47, XXY (Klinefelter) karyotype. While Klinefelter syndrome pregnancies are acceptable to some parents, Down's syndrome ones usually are not. Also, placental rather than fetal mosaicism is found to be more common in CVS preparations. This was reported by a number of investigators at the Rapallo meeting and subsequently published.[4] Mosaicism in CVS should be confirmed by amniocentesis before a pregnancy termination is recommended. This puts a bigger strain on the mother but avoids termination of normal fetuses.

BACK-UP FACILITIES

In centres where prenatal diagnosis evolved from already existing facilities, prenatal and other family studies can easily be carried out. In unexpected translocations and rearrangements, studies on parental chromosomes are vital, and sometimes provide a means for the correct interpretations of the fetal karyotype. An example is when a small metacentric fragment is found in the fetal karyotype. Fragments of this kind can be associated with multiple abnormalities.[9] But if such a fragment is also found in parental blood and the parent is apparently normal phenotypically, the pregnancy should continue. This was the policy adopted in our department, where 2 cases of non-inherited fragments were detected; the pregnancies continued and normal babies were born. Our general policy is that, unless experience tells us that a specific variation or rearrangement is related to an abnormal phenotype in a new born, termination of such a pregnancy is not recommended.

FOLLOW-UP FACILITIES

All centres providing prenatal diagnosis should have complete follow-up facilities for patients who delivered live born children, or who lost the pregnancy through stillbirth, abortion or selective termination. Normal deliveries can be notified by postal returns, but in cases where the mother lost her pregnancy such follow-up can be combined with supportive action. Borg and Lasker[1] stated that mothers confronted with loss of a pregnancy experience the same kind of grief as mothers who lose a child. Mothers also continue to ask themselves why the pregnancy was lost. If we can help to provide the answers, the process of healing can be faster. In cases of genetic terminations we have a moral and legal obligation to confirm the initial diagnosis on the product of conception.

CONFIDENTIALITY OF RECORDS

Information, sometimes very personal, has to be obtained to provide proper help and diagnosis. There are clear legal guidelines regarding the confidentiality of such records, and it is the responsibility of each centre to ensure the existence of such arrangements.

THE RESPONSIBILITIES AND PRIVILEGES OF THE SCIENTIST

They are quite simple. He or she must be totally honest about the limitations of the techniques and the laboratory's facilities. There must be complete and constant communication between the clinician and the scientist.

Prenatal diagnosis deals with the life and death of an unborn child; therefore it must not be abused for non-medical purposes like choosing the sex of the baby, or be reduced to a commercial commodity. If prenatal diagnosis is to remain a medical diagnostic procedure it must be coupled with human compassion and the highest ethical standards and integrity. Therefore, the scientist must retain the privilege of refusing to disclose the sex of the fetus if it is obvious that such knowledge will be abused.

BIBLIOGRAPHY

1 Borg, S. and Lasker, J.: 1982, *When Pregnancy Fails: Coping with Miscarriage, Stillbirth and Infant Deaths*, Rutlege and Kegan Paul, London.
2 Ferguson-Smith, M. A. and Ferguson-Smith, M. E.: 1983, *Problems in Prenatal Diagnosis. Contributions to the Eugenics Society Symposium on Developments in Human Embryology*, Academic Press, London.
3 Ferguson-Smith, M. E.: 1984, *Prenatal diagnosis: Cytogenetic problems and future developments*, Genetica Palonica, *25*, 289.
4 Fraccaro, M., Simoni, G. and Brambati, B., eds.: 1985, *First Trimester Fetal Diagnosis*, Springer-Verlag, Berlin.
5 Hamerton, J. L. and Ferguson-Smith, M. A., eds.: 1984, *Collaborative Studies in Prenatal Diagnosis of Chromosome Aberrations*, Prenat. Diag., Vol. 4.
6 Hsu, L. Y. F. and Perlis, T. E.: 1984, *United States survey on chromosome mosaicism and pseudomosaicism in prenatal diagnosis*, In, J. L. Hamerton, and M. A. Ferguson-Smith, eds., Collaborative Studies in Prenatal Diagnosis of Chromosome Aberrations. Prenat. Diag., *4*, 97.
7 Milunsky, A. and Amos, J. J., eds.: 1980, *Genetics and the Law II*, Plenum Press, New York.
8 Simoni, G., Brambati, B., Danesino, C., Rossella, F., Terzoli, G. L., Ferrari, M. and Fraccaro, L. M.: 1983, *Efficient direct chromosome analysis and enzyme determinations from chorionic villi samples in first trimester of pregnancy*, Hum. Genet., *63*, 349.
9 Tanheroni, W., Cao, A., and Furbetta, M.:1973, *Multiple anomalies associated with extra small metacentric chromosome: modified Giemsa stain results*, Hum. Genet., *18*, 291.
10 United Kingdom External Quality Assessment Scheme (UKEQAS): 1984/85, Association of Clinical Cytogeneticists, United Kingdom.

A. ZWINGER

FETOSCOPY AND CHORIONIC VILLI SAMPLING IN PRENATAL DIAGNOSIS OF GENETIC DISORDERS

Fetoscopy in the second trimester of pregnancy and chorionic villi sampling (CVS) in the first trimester are currently used in the Centre for Prenatal Diagnosis of Congenital Malformations and Genetic Diseases in the Institute for the Care of Mother and Child, Prague.

Fetoscopy for fetal visualisation was introduced in 1977, and fetal blood sampling, guided by fetoscope, in 1982. This relatively late application of the latter was because haemoglobinopathies constitute an extremely rare indication in Czechoslovakia. We use fetoscopy for sampling fetal blood in the prenatal diagnosis of haemophilia-A, for fetal intraperitoneal or direct intravascular blood transfusion in severe Rhesus isoimmunisation, and for fetal skin biopsy. Since 1984, we have also occasionally used the fetoscope for purposes of chorionic biopsy; usually, however, chorionic villi are aspirated by means of the PORTEX cannula, guided by ultrasound. Under fetoscope guidance we have also performed selective fetocide in one of a pair of twins found to have trisomy 21.

FETAL VISUALISATION

Fetoscopy for fetal visualisation is performed in the 17th-18th week of pregnancy using a standard fetoscope (manufactured by Richard Wolf, Knittlingen, West Germany), with a 2.2 and 2.7 mm diameter, following ultrasound localisation of the placenta, determination of the size of the amniotic sac, and the position and viability of the fetus. With anterior localisation of the placenta (in more than 40% of cases), the trocar of the fetoscope is introduced into the amniotic sac from the posterior wall of the uterus. This procedure was developed at our Centre. Visualisation of the fetus usually takes only a few minutes, but very occasionally up to 20–30 minutes. When a fetus is found to be affected, the pregnancy is terminated. Rhesus negative patients are always treated with human immunoglobulin anti-Rh O /D/.

By June 30, 1985, we had performed a total of 151 fetoscopies for fetal visualisation. This is the largest series of fetoscopies for this purpose registered by the International Fetoscopy Group. Table 1 summarises the indications and findings. These are divided into 4 groups. A more detailed listing of the first 3 groups is given in Tables 2–4. In 7 patients with genetic risk, fetoscopy was performed in subsequent pregnancies (in 1 case thrice).

The complications in the 151 fetoscopies for fetal visualisation are given in Table 5. With adequate skill and suitable arrangments, the risk of fetal loss should not exceed 5%. Intrauterine infection is the most frequent cause of spontaneous abortion. Obscured fetal visibility (2.6%) was generally caused by

39

E.K. Hicks and J.M. Berg (eds.), The Genetics of Mental Retardation, 39–44
© 1988 by Kluwer Academic Publishers.

TABLE 1

Indications for fetal visualisation by fetoscopy, and findings

Suspected condition	Number investigated	%	Fetus affected	%	Failed procedure	%
Digital/limb deformities (isolated or characterizing a multiple malformation syndrome)	57	37.7	8	14.0	–	–
Orofacial malformations (isolated or characterizing a multiple malformation syndrome)	60	39.7	18	30.0	1	1.7
Digital/limb and orofacial malformations (characterizing a multiple malformation syndrome)	25	16.6	4	16.0	–	–
Other malformations	9	6.0	2	22.2	1	11.1
Total	151	100	32	21.2	2	1.3

TABLE 2

Fetoscopy in pregnancies at risk for isolated digital/limb deformities or a characteristic multiple malformation syndrome

Suspected condition	Number investigated	Affected fetus
Ectrodactyly	17	3
Short rib polydactyly syndrome (Types II and III)	10	1
Holt-Oram syndrome	3	0
Peromelia	6	0
Syndactyly (Type V)	3	1
Achondrogenesis	3	1
Ectrocheiria	1	0
Syndactyly sec Haas	2	0
TAR syndrome	1	1
Adda syndrome	1	1
Diastrophic dwarfism	2	0
Bilateral equinovarus II	4	0
Freeman-Sheldon syndrome	1	0
Larsen syndrome	1	0
Familial brachydactyly	1	0
Oliver syndrome	1	0
Total	57	8

TABLE 3

Fetoscopy in pregnancies at risk for isolated or syndromal orofacial malformations

Suspected condition	Number investigated	Affected fetus
Familial cleft lip/platate	46	12
Treacher Collins syndrome	4	3
Van der Woude syndrome	5	1
Microtia type Schwarz III	1	0
Goldenhar syndrome (oculoauriculovertebral dysplasia)	2	0
Anencephaly	2	2
Total	60	18

TABLE 4

Fetoscopy of limb and orofacial malformations for prenatal detection of multiple malformation syndromes

Suspected condition	Number investigated	Affected fetus
Meckel-Gruber syndrome	7	0
Nager-De Reynier syndrome	2	0
Roberts syndrome	2	1
Fanconi's anaemia	2	0
Naumoff syndrome	2	0
Neu-Laxova syndrome	1	0
Smith-Lemli-Opitz syndrome	1	0
Apert-Pfeiffer syndrome	1	1
Robin syndrome (+ Equinovarus)	2	0
Ellis-van Creveld syndrome	3	1
Adam syndrome	1	0
EEC syndrome	1	1
Total	25	4

bleeding (even a small amount) from the uterine wall or, more often, from the edge of the placenta. In these cases, the procedure can be repeated about 2 weeks later, with visualisation improved by exchanging some 'cloudy' amniotic fluid with physiological saline. In our series, there was spontaneous abortion in nearly 4% of cases, compared with the combined world-wide experience in other centres of 7.7%.

TABLE 5

Complications of fetoscopic visualisation

	Number	%
Fetoscopy for fetal visualisation	151	100
Maternal injury	0	0
Fetal injury	1	0.6
Placental injury (with bleeding into the amniotic fluid)	2	1.3
Uterine wall bleeding into the amniotic fluid	2	1.3
Visualisation failure (at 1st examination)	4	2.6
Repeated puncture of the amniotic sac	3	2
Continuing pregnancies	119	100
Amniotic fluid leakage		
– early	7	5.8
– middle	3	2.5
– late	2	1.6
Intrauterine infection	4	3.3
Spontaneous abortion	5	4.2
Deliveries <37 weeks	10	8.4
Perinatal mortality	4	3.3

FETAL BLOOD SAMPLING AND SKIN BIOPSY

Fetal blood sampling from the insertion of the umbilical vessels is done by two different methods. In the first, a thin needle is introduced into the sampling site through a separate channel of the fetoscope. The latter has a 2.7 mm diameter and deflected optics for visualisation. A disadvantage of this method is that the narrow lumen of the needle makes blood sampling difficult. In the second method, a needlescope (Dyonics Inc., U.S.A.), with a 1.7 mm diameter, is introduced into the amniotic sac and a suitable site for blood sampling is visualised. Under needlescope guidance, a second puncture is made to introduce a needle into the insertion of the umbilical cord. The broader lumen of this needle facilitates blood aspiration; this outweighs the disadvantage of the procedure, i.e. the necessity for 2 punctures.

Fetal blood sampling was performed in 7 patients, and the amount of pure fetal blood aspirated was 0.8 ml. In all these cases, the indication was the risk for haemophilia-A in a male fetus; the fetal karyotype had been established previously from an amniotic fluid sample obtained by amniocentesis. Six fetuses showed normal values for factor VIII C. The remaining fetus was found to be affected and the pregnancy was terminated.

For fetal skin biopsy the site for sampling is found with the needlescope, usually in the gluteal region or on the thigh. The trocar cannula with the inserted needlescope is pressed to this site and the scope is replaced by a special fine biopsy forceps. We usually take 4 skin biopsies for electron-microscopy. The preferred period is the 20th-22nd week of pregnancy.

The indications for skin biopsy in 7 patients were risks to the fetus for one or other of the following skin diseases: harlequin ichthyosis (3 cases); epidermolysis bullosa lethalis (2 cases); lamellar ichthyosis, Coffin-Lowry syndrome (1 case each). Four affected pregnancies were found and abortions undertaken in each; the remaining 3 pregnancies resulted in the birth of healthy babies.

INTRAUTERINE BLOOD TRANSFUSION GUIDED BY FETOSCOPE

This is indicated in severe Rhesus isoimmunisation with a history of at least 1 previous fetus or child that died of erythroblastosis, and in the presence of Rhesus antibodies in the current pregnancy with amniotic fluid bilirubin at a dangerous level. Fetal hydrops (found at ultrasound) is a contraindication for intrauterine blood transfusion.

Intraperitoneal and intraumbilical venous transfusions, guided by the needlescope, with the transfusion administered through another insertion site, were performed by us between the 27th and 33rd week in 10 patients. In 2 patients the transfusion was repeated 3 times; in 4 cases, twice each; and in the remaining 4 patients, once each. The amount of Rhesus-negative blood administered was 30–100 ml in each case. We thus saved 5 out of 10 fetuses which, according to well-recognized criteria, would have had no chance of surviving.

FIRST TRIMESTER CHORIONIC VILLUS SAMPLING

First trimester CVS, proposed in China in 1975 for sex chromatin examination, was improved by adding ultrasonic guidance prior to aspiration.[1,2] Within a few years the combined world-wide experience on nearly 7,000 cases could be reported (at an international meeting in Giessen, Germany in 1985).

Analysis of chorionic villi cells enables fetal karyotyping, and diagnosis of dozens of enzyme deficiencies, and of diseases in which DNA defects are directly or indirectly involved.

CVS was introduced in Prague at our Centre in 1984. It is usually undertaken in the 9th-10th week of gestation by aspiration with the PORTEX cannula under untrasonic guidance. The needlescope is sometimes used for visually guided CVS with biopsy forceps. The quality of the sample is immediately checked under the dissecting microscope. We have been able to obtain sufficient chorionic villi at first attempt in over 80% of patients, though sometimes a second or even third attempt has been necessary. In our Institute the chorionic villi are used almost exclusively for direct karyotype analysis. Results are obtained within 2 hours of sampling.

At present, the risk of fetal loss after CVS is considerably higher than with amniocentesis (4–5%). Despite its advantages therefore, we offer CVS only to patients over 40 years of age, and to those at risk for an X-linked disease in a male fetus.

Fifty patients have had CVS in our Centre. Of these, 36 were studied because of advanced maternal age (> = > 40); 2 cases of trisomy 21 and 1 of trisomy 15 were diagnosed and aborted. The remaining 14 patients were at risk for an X-linked disease in the fetus; 6 male karyotypes were found and the pregnancies terminated. Thus, 41 continued, and 37 of these resulted in normal deliveries. The rate of spontaneous abortion after CVS was 9.7% (n = 4), but this should be reduced to 7.3%, as 1 case was a previously undetected blighted ovum with abortion 5 days post-CVS. Although this percentage is relatively high for CVS, we expect it to decrease as our experience increases.

BIBLIOGRAPHY

1 Brambati, B. and Simoni, G.: 1983, *Diagnosis of fetal trisomy 21 in first trimester*, Lancet, *1*, 586.
2 Gustavii,:1983, *First-trimester chromosomal analysis of chorionic villi obtained by direct vision technique*, Lancet, *2*, 507.

L. J. THOMASSEN-BREPOLS

PSYCHOSOCIAL ASPECTS OF PRENATAL DIAGNOSIS

In the past, the prevention of congenital disorders consisted primarily of magic rituals to deter bad spirits from harming the developing baby, and bringing offerings to the gods to solicit normal offspring. Indeed, the supernatural world was immanent in all things affecting health, livelihood and social activities. Our distant ancestors, believing in and seeking supernatural origins for most occurrences, were psychologically primed to expect the effectiveness of magic in curing the ill, preventing misfortune, and promoting fecundity.

Then, as the witch doctor/priest evolved into physician, pregnancy also came under the supervision of this young science of medicine. It initially facilitated control over conception, and thus the number of children fertile couples could have. In the last two decades, quality as well as quantity has become more controllable: selective abortion following prenatal diagnosis can prevent the birth of a seriously handicapped child; fetal treatment, although still distinctly limited, has also become increasingly possible.[29] As knowledge of human genetics increases, so also will the range of diagnosable conditions expand, which will undoubtedly have a profound impact on the psychological meaning of reproduction and pregnancy for human populations. The time when conception and pregnancy were considered acts of the gods is not so far behind us, and remnants of such beliefs may yet be part and parcel of our psychological make-up, albeit unconscious. Now, however, we must be psychologically prepared for the effectiveness of biomedical technology in the hands of fellow humans. Nevertheless, to fully grasp the psychological aspects of prenatal diagnosis, it can be helpful to be aware of these historical roots.

How does the public react to the possiblity of prenatal diagnosis? Is it, in fact, widely known? Do pregnant women generally accept and utilise this health service? Are the consumers satisfied with the procedure? What does pregnancy termination of an affected fetus mean to these consumers? The literature on these issues, concerned primarily with amniocentesis, is reviewed below.

UTILISATION OF AMNIOCENTESIS

Because population statistics on age specific birth rates are available in most western countries, maternal age is the only risk group for which utilisation rates can be computed. The total population of pregnant women at risk for other reasons (e.g., fetal neural tube defects, or inborn errors of metabolism) is not known.

Table 1 provides a summary of amniocentesis utilisation rates by older pregnant women for prenatal cytogenetic diagnosis. Apart from Denmark[46], where 56–72% of women over 34 years of age have prenatal cytogenetic diagno-

45

E.K. Hicks and J.M. Berg (eds.), The Genetics of Mental Retardation, 45–55
© 1988 *by Kluwer Academic Publishers.*

TABLE 1

Amniocentesis utilisation rates for prenatal cytogenetic diagnosis on basis of maternal age

Study reference	Place	Period	Lowest maternal age (years)	U-tilization rate
57, 59.	Netherlands	1976	38	6%
		1984		31%
7.	British Columbia, Canada	1976	38	8%
		1983		48%
9.	Australia	1979	40	29%
		1982		39%
10.	New York City, USA	1982	35	44%
25.	West of Scotland, UK	1976	35	8%
		1981		28%
51.	Upstate New York, USA	1979	35	23%
		1981		34%
30.	New York, USA	1979	35	29%
		1980		35%
46.	Copenhagen, Denmark	1980	35	72%
	Denmark	1980		56%
48.	Liverpool, UK	1979	35	16%
		1981		40%
40.	Montreal, Canada	1979	35	30%
35.	Paris, France	1979	40	28%
1.	Four states, USA	1978	35–40	6–28%
53.	Capetown, S. Africa	1977	35	10%
		1989		21%
54.	Georgia, USA	1978	40	15%
42.	Washington State, USA	1976	35	9%
		1977		13%
32.	New York, USA	1969	35	35%
		1979		45%
11.	Western New York, USA	1974–78	?	3%
5.	London, UK	1976	40	23%
		1979		48%
20.	Birmingham, UK	1977	40	16–26%

sis, utilisation rates range from low to moderate (3–48%), averaging out at 30% in recent years, with a slow general tendency to increase.

A puzzling finding in studies listed in Table 1 is the large regional variation in uptake of the service. Various factors in this regard are apparent. Thus, several reports demonstrate a marked influence of obstetrician's referral behaviour relative to the utilisation of this service. Bernhardt and Bannerman[12] found that 47% of New York obstetricians had never referred a patient for amniocentesis, and that those who did were most often younger members of the profession.

Religion is significant also, with fewer Catholic doctors offering the test.[11,12] Forster[28] noted that some doctors routinely discuss amniocentesis with their patients, while others give information about the test only if specifically asked. In Great Britain, for instance, different clinics apply different maternal age limits.[25] In Canada, 82% of non-tested women had not been given the relevant information by their doctors.[39,40] By contrast, 73% of women attending Upstate New York's regional programme received the information from their own obstetricians.[51] In short, there is a marked variation in the referral behaviour of physicians.

The wide acceptance of prenatal testing in Denmark is attributed[46] to the extensive discussion of the subject in newspapers, in women's magazines and on television, and to the centralisation of all prenatal care in hospitals where information dissemination about prenatal diagnosis is routine. Legalisation of pregnancy termination on demand is also widely approved.

ATTITUDES TOWARDS PRENATAL DIAGNOSIS AND ACCEPTANCE OF THE TEST

Though this varies, the public generally appears to have a favourable opinion about the option of prenatal diagnosis.[16,17] However, attitudes are not always the best predictors of actual behaviour, particularly when the issues are emotionally laden.[4] More informative are studies concerned with actual non-acceptance by eligible and knowledgeable pregnant women (see Table 2). The results indicate a wide variation in decisions, with 7–59% refusing the test option. Refusal appears to be lowest among well educated patients in private clinics[45], and highest among women from lower socio-economic groups.[43]

Previous history, e.g. earlier experience with a particular disorder, also plays an important role in a patient's attitude to prenatal testing.[21,31,36,44] Hsia et al.[31] in their study found that, after having had a child with a neural tube defect, most decided against a further pregnancy, and, of those who decided otherwise, 66% opted for prenatal testing. The availability of such testing can be the deciding factor for parents considering having further children.[22,31] However, Evans and Shaw[21] considered that, among carriers of haemophilia, "many seem to prefer to avoid the anxiety of having to think about the implications of prenatal testing by avoiding the possibliity of pregnancy entirely." The low acceptance of amniocentesis among this group has been linked to the possibility of self treatment for haemophilia patients, which allows them to lead a relatively normal life.[44] On the other hand, among Cypriot couples in Britain at risk for thalassaemia major who initially had no further offspring, 85–98% chose to have children when prenatal diagnosis became possible.[47]

The most important factor leading to refusal of prenatal testing appears to be the inability, on moral or religious grounds, to accept pregnancy termination (see Table 2). Other reasons include concerns related to the amniocentesis procedure, the risk of fetal injury or miscarriage, acceptance of handicapped life, disincli-

TABLE 2

Non-acceptance of amniocentesis by eligible women

Study reference	Place	Risk factors	Non-acceptance	N	Comments
10.	New York City, USA	all	28%	2027	high acceptance when service provision and information is good and there are no financial barriers
2.	Atlanta, Georgia, USA	NTD	37%	43	non-acceptance by blacks much higher: 71% (5 of 7), but the sample is small
51.	Upstate New York, USA	maternal age	40–50%	410	barriers: emotional and moral reluctance towards abortion; fear of risk to fetus
25.	West of Scotland, UK	maternal age	7%	?	moral and religious objections; 16% too late for referral; 50% not informed by doctor
58.	Rotterdam, Netherlands	maternal age	41%	69	25% anti-abortion, and accepting of handicap; 9% anxious; 7% dissuaded by doctor
40.	Montreal, Canada	maternal age	47%	66	36% consider test unnecessary; only 7% anti-abortion
45.	Philadelphia, USA	all	11–46%	209	acceptance related to source of referral; highest at private clinics
24.	Boston, USA	all	23%	316	fear of miscarriage; anti-abortion
43.	Atlanta, Georgia, USA	maternal age	43%	120	reasons against: anti-abortion and accepting of handicap; fear of procedure; opposition by husband/father; unnecessary; not wanting to know fetal condition
55.	New York, USA	all	51%	100	35% religious objections; 8% not wanting interference with nature; 8% unnecessary
19.	Sheffield, UK	maternal age	36%	148	older women (> 39 years) less non-acceptance than younger women (35–39 years)
56.	Nronx, New York, USA	all	59%	843	acceptance related to socioeconomic class and race

nation to interfere with nature, self-denial of the possibility of having a defective baby, and coercion by others (e.g., family, husband) against the test. Sometimes women who initially request amniocentesis later decline it on their doctor's advice.[57,58]

The general view among women who have had amniocentesis, and where no fetal abnormality was detected, has been very favourable. They would repeat the procedure in the future and would recommend it to others.[6,23,26,37,50,60] This positive attitude prevailed despite the considerable impact of prenatal testing, such as tension and anxiety during the waiting period (because of fear of fetal injury, miscarriage, pain or diagnostic error), and the worry of decision-making about the pregnancy should the fetus prove to be abnormal. Silvestre and Fresco[52] noted that the invasion of privacy by doctors and their technology ("medicalisation") was also considered unpleasant. Significantly more hostility has been observed in women who were to be tested than in those who did not require testing.[23] Such hostility may be due to anger at the prospect of reproductive failure while other women could remain carefree and enjoy their pregnancy. After test results were made known, hostility disappeared. Most women in the study referred to stated that prenatal diagnosis, and reassuring test results, had permitted their enjoyment of the second half of the pregnancy.

Very little is known about the attitudes to amniocentesis of women who had received a diagnosis of fetal abnormality and opted for termination of pregnancy. In a Dutch study[57], only 49% of 110 couples wanted another pregnancy after selective abortion. In subsequent pregnancies, all opted for repeat prenatal testing. The couples who refrained from further procreation did so mainly because of not wanting to face the possibility of another pregnancy termination.

PSYCHOLOGICAL ASPECTS OF PREGNANCY TERMINATION FOR FETAL DEFECTS

In contrast to the positive psychological reactions of most women following abortion for social indications, the sequelae of induced pregnancy loss as a result of prenatal cytogenetic diagnosis are frequently described as psychologically traumatic (see Table 3).

Why are these psychological reactions so severe? Thomassen-Brepols[57] systematically analysed 30 in-depth interviews with women following selective pregnancy termination. Independent judges rated these tape recorded sessions. For most of the women (77%), the event involved the loss of a wanted child, complicated by the fact that prenatal attachment (maternal-fetal bonding) had already occurred. Such loss implies mourning. Moreover, coping was further complicated by other losses which also required attention, e.g. loss of a feeling of biological and moral competence. Carrying a handicapped child was perceived as a failure to reproduce adequately. There was also a sense of moral incompetence through awareness of a personal contribution to the pregnancy loss. The loss did not happen to the couple passively, as is the case with a miscarriage,

50

TABLE 3
Follow-up of pregnancy termination after fetal diagnosis

Place study period study reference	Design	Research method	Number of subjects (N) and response rate (resp.)	Abortion interview interval in months	Abortion method	Results
New Jersey 1976–79 3.	ab-cohort	interview by questionnaire	N = 12 resp. 80%	3–33 a = 15.4	?	post-abortion depression; traumatic experience
California <1974 13.	ab-cohort	interview, MMPI	N = 13 resp. 87%	0–37 a = 21	?	92% depresses
UK 1976–79 18.	ab-cohort	unstructured interview	N = 12 resp. 80%	3–49 a = 17.7	hysterotomy = 7 prostaglandin = 5	high risk for depression and social disruption
Netherlands 1976–9 38.	ab-cohort	semi-structured interview	N = 19 resp. 95%	10–56 a = 32	prostaglandin = 10 hypotonic saline = 7 hysterotomy = 2	long-term psychological side-effects
UK 1977–81 41.	case-control	structured interview	N = 48 resp.?	?	?	77% had acute mourning symptoms resembling stillbirth reactions and unlike those with miscarriage or therapeutic abortion
Wisconsin ? 8.	ab-cohort	questionnaire	N = 45 resp. 63%	?	?	seeing, holding baby helped over 90% of couples in coping
California <1984 33.	ab-cohort	audiotaped interview (inter-raters)	f = 14 m = 12 N = 26 resp. 52%	4–43 a = 17.6	prostaglandin hypotonic saline	short term: adverse symptoms; long term: coping well; high non-response rate
Netherlands 1984 57.	ab-cohort	audiotaped interview (inter-raters)	N = 30 resp. 100%	6–27 a = 11	prostaglandin	43% had serious coping problems; 20% with still uncertain outcome; 33% coped well after initial depression

ab-cohort = group of women who underwent abortion for genetic reasons
MMPI   = Minnesota Multiphasic Personality Inventory

but involved their active decision to abort. Moral pain was inevitable. The choice was against life or, for a child, suffering. Loss of hope for a healthy child was another important factor, since a subsequent pregnancy would involve a similar or greater risk of fetal malformation. Finally, the couples frequently became socially isolated because people tend to avoid discussing problems of heredity, handicap and abortion. Complicating matters even further were the emotions elicited by two conflicting images, that of the desired conceptualised baby and the image of the handicapped child. Rationally, there is contentment with the prevention of the birth of a handicapped child, but emotionally there is mourning for the desired baby that will now not be born.[57]

These findings underscore pioneering observations by others.[13,14,27,34] It is known that coping is impeded by paradox, conflict and ambivalence, all of which appear to be amply present following pregnancy termination as a result of fetal disorders. Mourning might not succeed because couples fail to unravel the complexities of their situation, so that depression lingers on.

PSYCHOLOGICAL IMPACT OF FUTURE DEVELOPMENTS IN PRENATAL DIAGNOSIS

The development of chorion villi sampling (CVS) allows prenatal diagnosis to be made much earlier in pregnancy. For those chosing abortion, a safe, first trimester procedure can be performed. The psychological advantages, compared to amniocentesis, seem obvious. Nevertheless, as Perry et al.[49] pointed out, the risk of spontaneous abortion following CVS is decisive in a choice between amniocentesis and CVS. If this was stipulated to be 5% more for CVS, 71% would prefer amniocentesis; If the risks were equal, 82% would choose CVS. The actual risk has not yet been precisely established, but recent data seem promising.[15]

The ever widening range of detectable diseases and malformations will include relatively minor defects, genetic disorders which manifest later in life (e.g., Huntington's disease), and disorders having a wide range of symptoms, from hardly noticeable to life threatening. Even genetic predisposition for certain diseases (e.g., cancer or coronary artery disease), could be recognizable prenatally in the future. The psychological burden of decision making is thus likely to increase considerably, since it is already clear that less severe diseases and uncertainty about the degree of potential handicap (e.g. in some sex chromosome aberrations), pose the most problems for couples facing positive diagnosis and hence a decision concerning pregnancy termination.[27,38,57] CVS does not remove or neutralise the 'moral shape' of choices, even though it probably facilitates rationalisation, in that emotional attachment to the unborn might not be as strong as it is in the second trimester.

Reproductive decision making is further complicated should the fetus have the potential to become a patient who can not only be diagnosed, but also treated. Such potential brings new meaning to prenatal diagnosis. Questions which arise are: can the fetal malformation be corrected; to what extent; at what maternal and fetal risk? Perhaps we should be more aware that parental autonomy can be frightening. We might ask ourselves how to equip prospective parents so that they will not collapse under the psychological burden of making reproductive decisions.

CONCLUSIONS

Knowledge and acceptability of prenatal testing should yield high utilisation rates. Nevertheless, except for Denmark, these are modest. Referral practices of physicians and acceptance by women vary considerably. Unwillingness to accept pregnancy termination is the primary obstacle. Pregnancy termination after amniocentesis is a traumatic experience because it not only constitutes the loss of a wanted baby, but also involves other deeply felt symbolic losses, and thus engenders paradox, conflict and ambivalence. In reproductive decision processes, the subjective appraisal of the severity of the disorder appears to be a major factor. Future developments can hamper these processes as diagnosable conditions will include additional less severe, more variable diseases, and even correctable disorders. On the other hand, CVS may facilitate decision making if prospective parents feel less emotionally involved with the fetus at such an early stage in pregnancy.

BIBLIOGRAPHY

1 Adams, M. M., Finley, S., Hansen, H., Jahiel, R., Oakley Jr., G. P., Sanger, W., Wells, G. and Wertelecki, W.: 1981, *Utilization of prenatal genetic diagnosis in women 35 years of age and older in the United States, 1977 to 1978*, Amer. J. Obstet. Gynec., *59*, 673.

2 Adams, M. M., Weis, P., Oakley Jr., G. P. and Falek, A.: 1984, *Use of prenatal diagnosis among parents of infants with spina bifida in Atlanta, Georgia, 1976–1979*, Amer. J. Obstet. Gynec., *148*, 749.

3 Adler, B. and Kushnick, T.: 1982, *Genetic counseling in prenatally diagnosed trisomy 18 and 21: psychosocial aspects*, Pediatrics, *69*, 94.

4 Ajzen, I. and Fishbein, M.: 1980, *Understanding Attitudes and Predicting Social Behavior*, Prentice Hall, Englewood Cliffs, N. J.

5 Alberman, E., Berry A. C. and Polani, P. E.: 1979, *Planning an amniocentesis service for Down syndrome,*, Lancet, *1*, 50.

6 Ashery, R. S.: 1976, *A study of the relationship between manifest anxiety and social work intervention of couples having amniocentesis for prenatal diagnosis of genetic defects*, Studies in Social Work No. 110, Diss. Abstr., *36* (6), 7645-A.

7 Baird, P. A., Sadovnick, A. D. and McGillivray, B. C.: 1985, *Temporal changes in the utilization of amniocentesis for prenatal diagnosis by women of advanced maternal age, 1976–1983*, Prenat. Diag., *5*, 191.

8 Becker, J., Glinski, L. and Laxova, R.: 1984, *Long-term emotional impact of 2nd trimester pregnancy termination after detection of fetal abnormality*, Amer. J. Hum. Genet., *36*, 122S.

9 Bell, J. A., Pearn, J., Cohen, G., Ford, J., Halliday, J., Martin, N., Mulcahy, M., Purvis-Smith, S. and Sutherland, G.: 1985, *Utilization of prenatal cytogenetic diagnosis in women of advanced maternal age in Australia, 1979–1982*, Prenat. Diag., *5*, 53.

10 Benn, P. A., Hsu, L. Y. F., Carlson, A. and Tannenbaum, H. L.: 1985, *The centralized prenatal genetics screening program of New York City III: the first 7000 cases*, Amer. J. Med. Genet., *20*, 369.

11 Bernhardt, B. A. and Bannerman, R. M.: 1980, *Factors influencing the utilization of amniocentesis,* Amer. J. Hum. Genet., *32,* 98.
12 Bernhardt, B. A. and Bannerman, R. M.: 1982, *The influence of obstetricians on the utilization of amniocentesis,* Prenat. Diag., *2,* 115.
13 Blumberg, B. D.: 1984, *The emotional implications of prenatal diagnosis,* In, A. E. H. Emery and I. M. Pullen, eds., Psychological Aspects of Genetic Counseling, Academic Press, London, 201–217.
14 Blumberg, B. D., Golbus, M. C. and Hanson, K.: 1975, *The psychological sequelae of abortion performed for a genetic indication,* Amer. J. Obstet. Gynec., *122,* 799.
15 Brambati, B., Simoni, G., Danesino, C., Oldrini, A., Ferrazzi, E., Romitti, G., Terzoli, G., Rossella, F., Ferrari, M. and Fraccaro, M.: 1985, *First trimester fetal diagnosis of genetic disorders: clinical evaluation of 250 cases,* J. Med. Genet., *22,* 92.
16 Bundey, S.: 1978, *Attitudes of 40-year old college graduates towards amniocentesis,* Brit. Med. J., *2,* 1475.
17 Doherty, R. A. and Roghmann, K. J.: 1979, *Knowledge, attitudes and acceptance of prenatal diagnosis among women and physicians in the Rochester Region,* In, I. H. Porter, and E. B. Hook, eds., Service and Education in Medical Genetics, Academic Press, New York, 323–334.
18 Donnai, P., Charles, N. and Harris, R.: 1981, *Attitudes of patients after genetic termination of pregnancy,* Brit. Med. J., *282,* 621.
19 Duncan, S. L. B.: 1978, *Problems of prenatal screening program for Down syndrome in older women,* J. Biosoc. Sci., *10,* 141.
20 Edwards, J. H. and Webb, R.: 1979, *Amniocentesis rates in older women,* Lancet, *2,* 1301.
21 Evans, D. I. K. and Shaw, A.: 1979, *Attitudes of haemophilia carriers to fetoscopy and amniocentesis,* Lancet, *2,* 1371.
22 Evers-Kieboom, G., Frijns, J. P. and Berghe van den, H.: 1982, *Genetic counseling en prenatale diagnose na de geboorte van een kind met spina bifida of anencephalie het effect of de verdere gezinsplanning,* Gezond. Samenleving, *3,* 79.
23 Fava, G. A., Trombini, C., Michelacci, L. Linder, J. R., Pathak, D. and Bovicelli, L.: 1982, *Hostility in women before and after amniocentesis,* J. Reprod. Med., *28,* 31.
24 Feingold, M. and Crooks, W.: 1980, *Genetic counseling and decision making concerning amniocentesis,* Pediat. Res., *14,* 521.
25 Ferguson-Smith, M. A.: 1983, *Prenatal chromosome analysis and its impact on the birth incidence of chromosome disorders,* Brit. Med. Bull., *39,* 355.
26 Finley, S. C., Varner, P. D., Vinson, P. C. and Finley, W. H.: 1977, *Participants reaction to amniocentesis and prenatal genetic studies,* J. Amer. Med. Ass., *238,* 2377.
27 Fletcher, J. C.: 1973, *Parents in genetic counseling: The moral shape of decision-making,* In, B. Hilton, D. Callahan, M. Harris, P. Condliffe, and B. Berkley, eds., Ethical Issues in Human Genetics: Genetic Counseling and the Use of Genetic Knowledge, Plenum Press, New York, 301–327.
28 Forster, D. P.: 1977, *A survey of obstetric practice in the Trent region in relation to the prevention of Down syndrome,* Hlth. Lond. Publ., *91,* 179.
29 Harrison, M. R., Golbus, M. S. and Filly, R. A.: 1984, *The Unborn Patient. Prenatal Diagnosis and Treatment,* Grune and Stratton, Orlando, 1–139.
30 Hook, E. G. and Schreinemacher, D. M.: 1980, *Trends in utilization of prenatal cytogenetic diagnosis by New York State residents in 1979 and 1980,* Amer. J. Pub. Hlth., *73,* 198.
31 Hsia, Y. E., Leung, F. and Carter, L. L.: 1979, *Attitudes toward amniocentesis: surveys of families with spina bifida children, 1974, 1977,* In, I. H. Porter, and E. B. Hook, eds., Service and Education in Medical Genetics, Academic Press, New York, 303–321.
32 Jahiel, R. I. and Hansen, H.:1980, *Utilization of prenatal chromosome diagnosis in New York City. 1969–1979: underrepresentation of older women and clinic patients,* Amer. J. Hum. Genet., *32,* 112A.
33 Jones, O. W., Penn, N. E., Shuchter, S., Stafford, C. A., Richards, T., Kernahan, C., Gutierrez, J., Cherkin, P., Reinsch, S. and Dixson, B.: 1984, *Parental response to mid-trimester therapeutic abortion following amniocentesis,* Prenat. Diag., *4,* 249.
34 Kessler, S., ed.: 1979, *Genetic Counseling: Psychological Dimensions,* Academic Press, New York.
35 Lamy, M. L. and Leridon, H.: 1981, *Les grossesses tardives donnant lieu a une amniocentese,* Population, *1,* 174.

54

36 Laurence, K. M. and Morris, J.: 1981, *The effect of the introduction of prenatal diagnosis on the reproductive history of women at increased risk from neural tube defects,* Prenat. Diag., *1*, 51.

37 Leonardi, D. and Esrig, S. M.: 1982, *Concerns of patients before and after amniocentesis,* Amer. J. Hum. Genet., *34*, 99A.

38 Leschot, N. J., Verjaal, M. and Treffers, P. E.: 1982, *Therapeutic abortion on genetic indications; a detailed follow-up study of 20 patients,* J. Psychosom. Obstet. Gynec., *1*, 47.

39 Lippman-Hand, A. and Cohen, D. I.: 1980, *Influence of obstetricians' attitude on their use of prenatal diagnosis for the detection of Down syndrome,* Canad. Med. Ass. J., *122*, 1381.

40 Lippman-Hand, A. and Piper, M.: 1981, *Prenatal diagnosis for the detection of Down syndrome: why are so few eligible women tested?* Prenat. Diag., *1*, 249.

41 Lloyd, J. and Laurence, K. M.: 1985, *Sequelae and support after termination of pregnancy for fetal malformation,* Brit. Med. J., *290*, 907.

42 Luthy, D., Emanuel, I., Hoehn, H., Hall, J. B. and Powers, E. K.: 1980, *Prenatal genetic diagnosis and elective abortion in women over 35: utilization and relative impact on the birth prevalence of Down syndrome in Washington State,* Amer. J. Med. Genet. *7*, 375.

43 Marion, J. P., Kazzan, G., Fernhoff, P. M., Brantley, K. E., Carroll, L., Zacharias, J., Klein, L., Priest, J. H. and Elsas, L. J.: 1980, *Acceptance of amniocentesis by low-income patients in an urban hospital,* Amer. J. Obstet. Gynec., *138*, 11.

44 Markova, I., Forbes, C. D. and Inwood, M.: 1984, *The consumers' view of genetic counseling of hemophilia,* Amer. J. Med. Genet., *17*, 741.

45 McDonnell, A. E. and Zachai, E.: 1982, *Evaluation of referral source and family history in acceptance of amniocentesis,* Amer. J. Hum. Genet., *34*, 101A.

46 Mikkelsen, M., Fisher, G., Hansen, J., Pilgaard, B. and Nielsen, J.: 1983, *The impact of legal termination of pregnancy and of prenatal diagnosis on the birth prevalence of Down syndrome in Denmark,* Amer. J. Hum. Genet., *47*, 123.

47 Modell, B.: 1982, *Social aspects of prenatal monitoring for genetic disease,* In, H. Galjaard, ed., The Future of Prenatal Diagnosis, Churchill Livingstone, Edinburgh, 146–159.

48 Owens J. R., Harris, F., Walker, S., McAllister, E. and West, L.: 1983, *The incidence of Down's syndrome over a 19-year period with special reference to maternal age,* J. Med. Genet., *20*, 90.

49 Perry, T. B., Vekemans, M. J. J., Lippman, A., Hamilton, E. and Fournier, P. Jr.: 1985, *Chorionic villi sampling: clinical experience, immediate complications, and patient attitudes,* Amer. J. Obstet. Gynec., *151*, 161.

50 Rice, N. and Doherty, R.: 1982, *Reflections on prenatal diagnosis: the consumer's view,* Soc. Work Hlth. Care, *8*, 47.

51 Roghman, J., Doherty, R., Robinson, J. L., Nitzkin J. L. and Sell, R. R.: 1983, *The selective utilization of prenatal genetic diagnosis: Experiences of a regional program in Upstate New York during the 1970s,* Med. Care, *21*, 1111.

52 Silvestre, D. and Fresco, N.: 1980, *Reactions to prenatal diagnosis: an analysis of 87 interviews,* Amer. J. Orthopsychiat., *50*, 610.

53 Smart, R. D.: 1981, *Down syndrome in the Cape Peninsula and the value of amniocentesis as a preventive measure,* S. Afr. Med. J., *59*, 670.

54 Sokal, D. C., Rogers Byrd J., Chen, A. T. L., Goldberg, M. F., and Oakley, Jr. G. P.: 1980, *Prenatal chromosomal diagnosis: racial and geographic variation for older women in Georgia,* J. Amer. Med. Ass., *244*, 1355.

55 Susca, L. A.: 1980, *Factors affecting decisions of pregnant women who received detailed instructions concerning availability of amniocentesis,* Amer. J. Hum. Genet., *32*, 131A.

56 Taben, E. and Nitowski, H. M.: 1978, *Non-compliance following referral for midtrimester amniocentesis,* Amer. J. Hum. Genet., *30*, 69A.

57 Thomassen-Brepols, L. J.: 1985, *Psychosocial Aspects of Prenatal Diagnosis,* M.D. Thesis, Erasmus-universiteit, Rotterdam.

58 Thomassen-Brepols, L. J., Duivenvoorden, H. J., Verhage, F. and Galjaard, H.: 1983, *Prenataal onderzoek bij vrouwen van 38 jaar en ouder,* Med. Contact, *52*, 1637.

59 Thomassen-Brepols, L. J., Jahoda, M. G. J., Drogendijk, A. C. and Galjaard, H.: 1982, *De lage opkomst in Nederland van oudere aanstaande moeders voor prenatale diagnostiek*, Ned. Tijdschr. Geneeskd., *126*, 2262.
60 Verjaal, M., Leschot, N. J. and Trefffers, P. E.: 1982, *Women's experiences with second trimester prenatal diagnosis*, Prenat. Diag., *2*, 195.

K.-H. GUSTAVSON

ETHICAL QUESTIONS IN THE PRENATAL CHROMOSOMAL DIAGNOSIS OF MENTAL RETARDATION FROM THE PAEDIATRICIAN'S AND CLINICAL GENETICIST'S POINTS OF VIEW

The purpose of prenatal chromosomal diagnosis of mental retardation is to identify a specific chromosomal aberration of a severity which could be considered to justify termination of the pregnancy. Fortunately, in the great majority of these prenatal investigations, the fetus is found to have a normal chromosomal complement, which is, of course, very reassuring to the pregnant woman and the prospective father.

The parents must be given information about the investigation procedures, including the risks involved in the sampling itself. It is important that they should know, for example, that the sampling procedure can sometimes cause a miscarriage. It may be difficult for the parents to balance the risk of miscarriage of a healthy fetus as a result of the examination against the possibility of detection of a fetal abnormality. It is unlikely that all parents who consider having a prenatal chromosomal investigation performed will be interested in learning about all the abnormalities which it may reveal. Nor is it reasonable to assume that all parents will have decided from the outset to have the pregnancy terminated if the fetus has a chromosomal aberration. They cannot make this decision until such an aberration has actually been discovered, and they have learned what this involves. Before the investigation is performed, however, the parents should have considered their attitude towards possible termination of the pregnancy. They should also be informed that both serious and less serious chromosomal aberrations may be found, so that they have an idea as to what questions may arise as a result of the findings.

The parents should be given an opportunity to talk to a midwife or physician about the implications for them of a prenatal chromosomal analysis. Such a conversation can never be suitably replaced by an information pamphlet, although written information is also valuable. A combination of oral and written information is the most satisfactory. A discussion allows the parents to ask questions and express their feelings, and it also gives them contact with someone to whom they can turn if they want further information. A well-formulated pamphlet, provided beforehand, can serve as a basis for questions. The mothers can be given such a pamphlet at their first visit to the antenatal clinic, and asked to read it in their own time. Subsequently, if desired, the parents should have the

E.K. Hicks and J.M. Berg (eds.), The Genetics of Mental Retardation, 57–60
© 1988 *by Kluwer Academic Publishers.*

opportunity to discuss the possibility of prenatal investigation with the midwife or physician.

As with any other medical investigation, prenatal chromosomal analysis needs informed consent, and for this the woman (or couple) requires adequate information about the procedure, the risks involved in amniocentesis or chorionic villus biopsy, and the possible effects of a chromosomal aberration on the fetus.

It is important that the parents are not given the impression that a prenatal chromosomal investigation is a routine procedure to which they just have to say "no" if they do not wish to have it done. On the contrary, those parents who decide to have the chromosomal analysis performed should be aware that they have made an active choice.

All prospective parents want assurance that their expected child will be healthy. Prenatal chromosomal investigation can erroneously be regarded as a way of obtaining such a guarantee. Those parents whose notion is to to use the test in this way suppress thoughts that the result might produce a situation in which they are forced to consider termination of the pregnancy. Their reason for fetal diagnosis is an attempt to protect themselves from such thoughts. In these cases correct information is especially important. The parents should be told that certain, but far from all, congenital defects can be discovered prenatally, and that it is never possible to guarantee that the expected child will be perfectly healthy. Some parents only want to have certain circumscribed information. Perhaps they may want to know, for instance, whether the fetus has a chromosomal abnormality that will lead to Down's syndrome, but not the gender of the fetus, or whether there is any sex chromosome aberration. One should comply with such requests, which in themselves indicate that the parents are well-informed to some extent.

Even though the woman intends to terminate the pregnancy if the fetus should be found to have a severe chromosomal aberration, she has the right to alter her decision. The parents should be informed before the sampling that less serious or relatively benign chromosomal aberrations may also be found, e.g a sex chromosome deviation with a low risk of clinical abnormality. They should also be told in advance about the diagnostic limitations of the testing procedure.

Most couples who ask for a prenatal chromosomal investigation have a strong desire to have children, but nevertheless they are prepared to terminate the pregnancy if the fetus shows a serious chromosomal abnormality. For the majority of parents, information that the fetus has a chromosomal aberration is completely unexpected. Even though they have the prenatal investigation because they want to rule out this possibility, they have difficulty in imagining beforehand that this can happen to them. As in the case of other events with far-reaching adverse consequences, the information concerning the fetal abnormality leads to a traumatic crisis. Different people react differently to such a crisis, but the reaction usually follows a characteristic pattern. Almost invariably such questions as "why has this happened to me, now?" are asked. Generally there is no satisfactory answer which can be given.

Medical explanations are always incomplete. The individual giving the medical information should therefore be aware that it is often considered inadequate by the afflicted persons. It is common for parents who are told that their expected child has a chromosomal aberration to have feelings of guilt. Assurance that there is no reason for such feelings is often not convincing. The parents have to be able to work through their crisis and thereby alter their attitudes towards themselves, to their lives, and to what has happened. The finding of a chromosomal abnormality affects the parents' confidence in their ability to have healthy children in the future. This is where genetic counselling is valuable. Obviously it is for the parents themselves to decide whether they will try for a new pregnancy. Some of them will do this fairly soon, while others will opt not to have another child. The counsellor should not encourage them in either direction.

There are two occasions in particular when the parents need to be given information and support:

1. When they are trying to decide whether or not to have a prenatal investigation performed, and

2. When an investigation has revealed that the fetus has a chromosomal aberration and they have to decide whether the pregnancy should be allowed to continue.

Both the sampling and an abortion concern the woman's body. She therefore has to ultimately decide whether a prenatal investigation or an abortion should be carried out. As far as possible however, the father should participate in this decision; otherwise there is a risk that he will distance himself from decisions that are of major importance for the couple's future life together. The father should therefore be given information in the same way and preferably at the same time as the mother. They will then have an opportunity to interact and arrive at joint decisions.

A selective termination of pregnancy has much in common with a stillbirth. The parents may therefore need help in their grief process. One way of giving them this is to offer them the opportunity, but obviously not to force them, to see the aborted fetus. This is possible with a late, but not an early, abortion. Even if it is painful for the parents to see the fetus, this can help them to get a realistic picture of the abnormality and give them something tangible to mourn. Another possiblity is to offer to show the parents a photograph of the fetus. Naturally, they have the right to know the results of pathological examination of the fetus, and should be given this information. It can be important for them to conclude that they were right in choosing to terminate the pregnancy.

Some parents who have chosen abortion want to end their relationship with the fetus by a funeral service or other religious ceremony. In some cases this is probably an expression of a legitimate parental need to deal with feelings of guilt towards that fetus. Irrespective of their attitude to a funeral service, certain parents who decide upon a selective abortion clearly have a great need to talk to a minister of religion.

Especially when a 'mild' chromosomal aberration has been discovered, the parents may want to have the child even though they know that it will not be 'completely' normal. This is relatively uncommon at present, but possibly the number of parents who make this choice will increase, as the development of increasingly sophisticated techniques lead to the discovery of more 'minor' fetal chromosomal aberrations. Parents who choose to give birth to a child with a chromosomal abnormality naturally feel that they have taken upon themselves a great responsibility. Some of them do not wish to be reminded of this by extra attention, but prefer that the pregnancy be treated as any other. Others may want special support and as much information as possible. It is important to be aware of the parents' wishes in these respects and to give them help on their own terms.

Talking to parents who have been told that their fetus has a chromosomal aberration is not merely a matter of providing medical information. It is, to a large extent, an occasion for responding to the parents' feelings. How this is done is greatly dependent on how the counsellor perceives his/her role, and on the counsellor's attitude towards fetal abnormalities.

J.C. FLETCHER

EVOLUTION OF ETHICAL ISSUES IN PRENATAL DIAGNOSIS

INTRODUCTION

This paper explores how ethical issues evolve in different societies in the practice of prenatal diagnosis of genetic disorders. Following a brief explanation of the author's view of ethics, a hypothesis of a two-stage process in the evolution of ethical issues in variable cultural settings is considered. These ideas were developed during a cross-cultural study of consensus and variation in approaches to problems of moral choice among medical geneticists in 19 nations.[a] A questionnaire about the ethical reasoning of medical geneticists facing typical cases involving problems of moral choice has been sent to more than 1,000 medical geneticists in these nations. The results will be analysed and reported in due time.

Ethics is the best thinking about the best interests of individuals, groups and society in the context of socio-moral conflicts. Human thought about the socio-moral conflicts and guidance regarding such conflicts is transmitted in the ethical traditions of a culture. In contemporary societies, philosophical or religious traditions are major sources of ethical thought. However, political, social and professional movements also add their contributions.

Clearly, a society applies selection to identify whose thinking is 'best' about the most significant moral conflicts. These conflicts arise from biological and cultural necessities and emerge as issues of life and death, war and peace, health and sickness, competition and cooperation, and other antinomies of life. In such socio-moral conflicts, individuals and groups must inevitably choose ways to envision their best interests and the best interests of those with whom they contend and strive. The social process of ethics is to become conscious on how interests are envisioned and to use reasoned arguments about choices in terms of societal, group, and individual interests.

Ethics is predominantly selected on the basis of biologically and culturally shaped self-interests of persons and groups. Like every other human activity, ethics is a limited social and historical enterprise. However, a margin of freedom does exist in human affairs by which human reason and foresight can gain perspective on the biological and cultural forces that shape what is deemed to be 'best.' If this premise is acceptable, and if one takes the long view, it is reasonable to examine the consequences of pursuing alternative moral visions and their approaches to moral problems. By careful evaluation of the consequences of consistently following one or more moral approaches to a socio-moral conflict, we can collectively look ahead to foresee what approach to the

61

E.K. Hicks and J.M. Berg (eds.), The Genetics of Mental Retardation, 61–72
© *1988 by Kluwer Academic Publishers.*

conflict will likely serve the best interests of most persons directly and indirectly affected by the particular conflict. In short, every system of ethics ought to be evaluated primarily by the consequences it will produce in society. This brief sketch of my view of ethics is more fully discussed elsewhere.[b]

Ethics, these socially-created bodies of best thinking, evolve in a process with three partners: morality, law, and science and technology. The distinction between morality and ethics should be noted. Moralities are culturally produced systems of rules, obligations and duties designed to resolve the most familiar conflicts of moral choice. Morality is what one learns by being 'well brought up' in a society. However, moralities differ from society to society. Indeed, even within the same society, competing moralities contend and clash. I was raised in the Southern United States and can testify that Southern morality is different in many respects from Northern morality. I spent 6 months in Norway in 1984 and know that Norwegian morality is different from Danish morality, and so on. Yet, it is possible to gain some distance on our learned or intuitive moralities and guide and even reform them when necessary, i.e., when loyalty to them would do more harm than good.

The best tools for ethical reflection on morality and its effects are critical ethical principles, such as autonomy, non- maleficence, beneficence and justice. Each of these principles has a history and a wide range of contemporary meaning and application. Familiarity with the history and relevance of critical ethical principles helps to gain perspective on the moralities that human beings inherit as part of their cultural package. By examination of the harm or good that will likely be done by loyalty to a specific moral approach to a problem, ethics is supposed to criticize and guide morality in the long run. I shall return to these particular principles in discussing cultural differences on moral choices in prenatal diagnosis.

Law is the process by which societies order their moral priorities and reward and punish loyalty or disloyalty to the moral order in the most direct way. Law is always stoically relevant, but often a poor tool for directly resolving complex choices presented by science and medicine.

Science and technology comprise all that is known and will yet be learned about the universe, as well as the tools by which this understanding is gained and managed. Science and technology are frequently the occasions for significant ethical issues but are not the sources of ethical guidance. However, lay persons 'on the street' tend to be determinists who believe that science and technology, rather than ethics, are the chariot drivers of this 3 horsepower scheme, rather than ethics. "If it can be done, it will be done" is the most frequent response when asked about the ethics of science and technology. I do not agree with this view. There are many things that could be done in science and medicine that are not done for ethical reasons. Fetal and genetic research are good examples of this restraint. However, societies differ in the degree of social freedom provided for serious ethical criticism of morality, law, science and technology. The social

enterprise of ethics can be defeated or corrupted in some societies by narrow ideological or nationalistic interests.

THE EVOLUTION OF ETHICAL ISSUES IN PRENATAL DIAGNOSIS: EARLY STAGE

My thesis is that ethical issues (socio-moral conflicts) have evolved in two stages in variable cultural contexts. At the early stage some key ethical issues are raised concurrently with introduction of prenatal diagnosis as an innovation in a society. Certain innovators, in this case physicians, patients and families, have the knowledge and courage to be the first generation to use the innovation. The study of diffusion of new technologies[16] in different cultures shows that respected innovators are the most important and consistent feature in the adoption of an innovation. Without innovators, who in this case are also agents of change in cultural evolution, prenatal diagnosis could not have been introduced in any society.

Prenatal diagnosis was introduced in the United States and other nations in the late 1960s in the midst of serious moral and ethical challenge, especially by those protective of fetal rights or concerned about adverse consequences for living and future handicapped persons. The physicians and ethicists who raised the earliest objections against prenatal diagnosis did not attack the technique itself or express bias against technology. They opposed selective abortion and offered various moral and legal reasons for their opposition which can be organized under two themes: (1) a basic purpose of medicine – to save life – is violated by the practice of abortion; and (2) that, while some abortions may be justified, the use of prenatal diagnosis tends to set apart certain fetuses as deserving of abortion and thus treats fetuses unequally and unjustly. On the other hand, the innovators argued that prenatal diagnosis and selective abortion grew out of (1) the obligation to reduce or prevent suffering for the affected family, the fetus in question, and society, and (2) the obligation to prevent genetic disease and its impact on future generations in the absence of successful genetic therapies.[c]

In the dilemmas faced by the innovators and those who initially challenged them, the earliest stage of evolution of the issues occurred. Gradually, a dominant moral approach was shaped, tested by experience, to guide medical geneticists and parents in their choices about the most difficult ethical problems in prenatal diagnosis. I summarize below this dominant approach after describing the five main ethical issues of the early stage, which are ranked in order of historical appearance and extent of consideration in the medical-ethical literature.

1. Difficulty in abortion choices was the first and most serious ethical issue to arise between physicians, parents and families, as well as between innovators and challengers. The most serious moral disagreements in an early (1970) international meeting of medical geneticists convened by the NIH was whether

amniocentesis ought to be provided at all to couples who would not undergo abortion.[6] Some centers required a pre- commitment to use abortion to receive prenatal diagnosis.[9] This was challenged by physicians and others who believed that more choice and respect for autonomy should prevail throughout the counselling process. This latter view gradually prevailed and early restrictive consent documents were set aside.

Abortion choices have been difficult in the context of prenatal diagnosis by amniocentesis because of (a) views of the higher moral status of the fetus at mid-trimester, (b) the wide spectrum of severity in some diagnosable genetic disorders (e.g., sickle cell disease), (c) the ability to treat some disorders, and (d) arguments by challengers that linked abortion at mid-trimester with infanticide in neonatal choices. Will first trimester chorionic villus sampling present fewer moral and emotional difficulties? Such comparative studies need to be done in the context of comparing the safety and accuracy of first and second trimester procedures.

2. Controversial indications for prenatal diagnosis was the second ethical issue. When organized genetic services were first offered, the indications in centers were largely directed towards well-defined categories of pregnancies known to be at higher risk.[13] However, questions soon arose whether prenatal diagnosis should be done apart from these criteria, or on request by a pregnant woman expressing "anxiety." This problem is pressing, especially where prenatal diagnosis is rationed by the health care system or the lack of widespread expertise. In Norway, the maternal age cut-off is 38 years. Amniocentesis is rationed by Parliament, which also requires that oversight of medical genetics be provided by a special committee accountable to the Parliament.[2] By contrast, in nations such as Switzerland and the United States, where market forces and consumer preferences play a larger role in the health care system, far less control is exercised over the indications policy. In Switzerland, of the 12,038 amnio-centeses done between 1981 and 1983, 3,175 (26.4%) were for women under 35 for the indication of 'anxiety.'[17]

Sex choice unrelated to a sex-linked disorder is another controversial request. Cases of unwanted gender arise mainly in families with several children of the same gender or in families from cultures with strong male preferences.[3,7] Objections to prenatal diagnosis for sex choice have been made to conserve resources, protect against gender discriminations, and to avoid a backlash from those who oppose abortion.[7,8,10] My own position adds a fourth reason. The problem of sex selection tends to reintroduce a feature of 'positive' eugenics into contemporary medical genetics. Negative eugenics involves the prevention of harms from genetic conditions. Positive eugenics is an attempt to improve the species by enhancement of existing genes or deliberate selection of genotypes for mating deemed to be superior. No genetic harm is involved in a family that has been unfortunate in natural assignment of gender at fertilization. To attempt to 'treat' their disappointment by prenatal diagnosis and abortion not only harms the fetus for a less weighty reason but may encourage ideas of deliberate preselection of

gender. Since the sex ratio is virtually perfect, except in societies that systematically neglect female children at birth, change in the sex ratio is not needed.

3. *Problems in disclosure of diagnostic results* also occurred, first due to learning the sex of the fetus and then due to disputed laboratory findings or results that were arguably harmful in a psychological sense to parents or the future child. A physician could legally refuse to disclose fetal sex to parents out of concern about abortion, since these data are usually incidental and not involved with disease. However, the practice is to disclose fetal sex to parents who request it. True versus pseudo-mosaicism or the possibility of contamination by maternal cells creates genuine doubt. When too much time has elapsed to undertake a repeat diagnosis, what should the parents be told? What if sonography reveals an irregularity of the fetal head but the amniotic fluid is normal for alpha-fetoprotein? Also, is the counselor obliged to disclose the history of controversy about criminality and XYY individuals?

4. *Twin pregnancies* present difficult dilemmas when one twin is diagnosed as having a disorder and the other is presumed normal. When this is discovered, parents have three choices: continuation to term with the higher risks of any twin pregnancy, abortion of both twins, or selective feticide of the affected twin in the interest of the selective birth of the healthy twin. The dilemma involves considerations of possible harm to the healthy twin and avoiding risks for the mother of the dangers of clotting, hemorrhage and shock. To date, none of these consequences has been reported in the literature on twin pregnancies and prenatal diagnosis.[1]

5. *Access to and distribution of services* is the final issue, perhaps most important in terms of persons directly affected. The problem is one of fairness. Understandably, when prenatal diagnosis was being introduced, only a small fraction of persons who might have benefitted were actually served. But many nations' pregnant women and families suffer greatly because access and distribution are unfairly arranged. Modell, among others, has consistently viewed access to and allocation of prenatal diagnosis as the central ethical problem in medical genetics. She and others reported that in most developed nations less than 50% of women over 35 years of age have prenatal diagnosis, although about 80% request it when informed.[11,12] Except for Denmark, where about 80% of women with medical indications receive prenatal diagnosis, no other nation meets the need for services. The best estimate of utilization in the United States is that probably no more than 25 to 30% of all pregnancies at known risk for genetic reasons are studied, although in some major urban areas utilization is 50% or higher.[13]

HYPOTHESES FOR RESEARCH

With the appearance of these ethical issues in the early to mid-1970s, the need for ethical guidance was increasingly recognized. Through efforts by the Hastings

Center (USA) and in international meetings of medical geneticists, the first moral approaches to these and other problems were fashioned.[5,14] In the United States, a President's Commission on Ethical Problems in Medicine roughly followed the same pathways charted by these earlier groups, although its report encompassed genetic counselling and screening as well as prenatal diagnosis.[15] A significant congruence of principles and moral approach is found in the guidance provided in these three sources. With generous support from my institution, the University of Oslo, and the Norwegian Marshall Fund, 24 genetic centers in 11 West and East European nations were visited in 1984 to study the question of whether the dominant moral approach found in these earlier statements was shared by medical geneticists in these sites. Also, a second question was framed as to whether such early guidance was sufficient for the ethical issues now arising for medical geneticists and society.

I hypothesized that a dominant moral approach to the early ethical issues had clearly emerged by 1984. With notable cultural differences, there were three features on which strong consensus existed among medical geneticists: respect for parental autonomy, non-infliction of harm on individuals and families, and voluntary programs of screening and genetics education.

In these site visits, I found a strong affinity between those features and the approaches of European medical geneticists, although clear cultural differences were also evident. Reliance on parental autonomy was seen to be the major source of guidance in conflicts about abortion choices, although this doctrine reached its limits in cases of prenatal diagnosis for sex choice. Non-infliction of harm by the prevention of genetic disease was seen to be the major justification for selective abortion, especially in cases of untreatable and serious genetic disorders. Emphasis on the voluntary basis of screening and education reflects loyalty to the autonomy principle as well as the intent to benefit individuals and families at higher genetic risk. Strong consensus against any type of coercion in the practice of prenatal diagnosis was found virtually everywhere.

Physicians and their experience with patients undoubtedly had the greatest influence in shaping this dominant approach, although they drew upon values and principles from a general cultural ethos that strongly resists coercive interference with reproductive plans and selection of mates. Although not without continuing ethical criticism of prenatal diagnosis, this dominant moral approach actually functions to guide decision making. More cultural variability can be found in the issues of full disclosure of controversial or potentially harmful results of prenatal diagnosis and about the justice issues in access and distribution of services. Wide variability also exists in serving women under 35 years old and parents who are opposed to abortion.

Site visiting and observation are preliminary and subjective methods of studying problems, including practical ethics. Therefore, a questionnaire was designed to test this hypothesis cross-culturally, in tandem with a second hypothesis that a newer and more complex stage of evolution of the ethical issues in prenatal diagnosis is already developing. The scientific and medical setting of medical genetics, well described in M. A. Ferguson-Smith's paper (this volume), is the source of rapidly developing technical trends in prenatal diagnosis and allied fields which have already launched a second and more complex stage in ethical issues. Have medical geneticists and their societies only just come to some agreements, many of them not without serious continuing dispute, only to be faced with a significantly new agenda of ethical issues for which very little established guidance exists? This appears to be more or less the case in several nations. The results of the questionnaires will inform us more accurately.

The second hypothesis is that technical trends will introduce new and more complex choices for medical geneticists, parents and policy makers that will increasingly involve society's interests. It is highly questionable whether simple reliance on the dominant moral approach sketched above will suffice to guide in the complexities of new issues: e.g., a) increasing demand for services, b) the potential for general pregnancy screening, c) the potential for predictive genetic screening of heterozygotes whose genes make them more susceptible to common diseases, d) wider carrier screening, e) the potential for sex selection and f) research to determine whether genetic diagnosis in the pre- implantation embryo is feasible for selection of embryos for implantation and, if successful, the avoidance of abortion for families at the highest genetic risk. Advances in genetic knowledge will also continue to raise issues about positive eugenics and society's ability to draw and maintain clear moral lines between treating real disease and trying to improve the species.

The main characteristics of the later stage of ethical issues in prenatal diagnosis are more complex interactions with society, multiple issues and implications, the need for new guidance beyond the dominant approach, and the fact that many more actors are involved in debate of the ethical and public policy issues.

HOW DO ETHICAL ISSUES EVOLVE IN DIFFERENT SOCIETIES?

The best way to study ethical problems in prenatal diagnosis in different societies is to invite those who are presumed to have the problems to identify them and to explain their reasoning behind the choices they routinely make. However, it is also necessary to shape a theory as to how ethical issues in prenatal diagnosis evolve in different societies. Further, it is necessary to decide how much weight ought to be given to cultural differences in ethics. Just because cultural differences exist, it does not necessarily follow that ethics is entirely relative to culture and personality. Are there universal principles which carry sufficient weight to guide difficult ethical choices despite cultural differences? Cross-cultural studies will address the question of whether there is any basis or common ground for expressions of national and international ethical consensus about the most significant ethical issues in the practice of medical genetics, and its relation to societies. Are there issues of moral choice upon which medical geneticists are so clearly agreed that they could adopt written guidelines?

At this point, the elements in a theory of evolution of ethical issues in prenatal diagnosis are being explored. We must learn first how the issues evolve at all, in order to search for deeper common ground in the realm of ethics. It appears that 4 cultural factors strongly influence ethical issues in prenatal diagnosis in different societies. These factors act as cultural biases that accelerate or decelerate the issues towards more or less complex resolutions.

First, in each society there are competing moralities that clash about the ethical justification for prenatal diagnosis, especially about abortion choices. If the dominant morality in the society permits elective abortion, the society will likely have adopted the dominant approach and be well along into the second stage, assuming that the technological capacity for development is present. If emphasis on parental autonomy and choice for abortion is expressed in a sub-dominant morality, and if the dominant morality condemns or strongly challenges the premises upon which abortion arguments are based, the elements of the dominant approach to the major ethical issues in the early stage will be incompletely formed or reflect a wide degree of variation and lack of consensus.

Secondly, if an abortion law exists that clearly sanctions abortion for genetic reasons and if public policy encourages the protection and privacy of those who seek prenatal diagnosis, it will be more likely that a later stage will be found.

Thirdly, the more highly developed the technological and scientific base of the health care system, and the more firmly established medical genetics is found to be in that system, the more will ethical issues be integrated into all levels of the social nexus.

Fourthly, if the level of concern of the society with the well- being of families is high, and if the social and economic status of women in the society is rising, much more evidence of later stage development of ethical issues will be found. The less supportive societies toward families and women will still be found to be in significant conflict in the earliest stage.

Another aspect of an interpretative theory of the evolution of ethical issues lies in the notion of a deeper backgound of the critical ethical principles and the ways that different societies bias selection and ordering of these ethical principles that have the most direct bearing on the best interests of those who provide and use prenatal diagnosis. Which principle ranks highest? In the United States, autonomy, or the respect for self- determination, ranks very high in medical ethics, sometimes to the detriment of society's interests. Justice ranks quite low. Americans tolerate significant inequalities in health care and education. If selection of principles were examined in the People's Republic of China, justice would rank first, then beneficence, non-maleficence and autonomy perhaps lower. Different societies select and order ethical principles in response to cultural biases and the imperatives of the prevailing economic and social system. Critical ethics serves to argue for a reordering of ethical principles. These universal ethical principles could be the source of common ground between medical geneticists from different nations and cultures upon which to examine their agreements and disagreements. The dominant moral approach to the major ethical issues in medical genetics is undoubtedly a product of Western, post-industrial societies. This approach may not be the most appropriate in less developed nations in the Far East, Mid-East and African continents. Also, the dominant moral approach may well need to be reformed, especially if mastery over genetic manipulation of the human genome is gained. We are in only the earliest stage of the moral evolution that must accompany and guide the scientific and technical aspects of medical genetics.

CONCLUSION

Is there a need for ethical consensus in medical genetics today? With Norwegian colleagues, three reasons to support an affirmative answer have been given.[4] First, older bodies of ethical guidance used by physicians in many nations need to be supplemented (not supplanted) by practical ethical statements based on the experience of practitioners of prenatal diagnosis. Many current problems in medical genetics call for skills in counseling and ethical sensitivities which exceed the demands of everyday medical practice. Medical geneticists deal daily with choices about human reproduction, an area of cultural life filled with strong emotions and laden with moral and ethical considerations. Consensus about key problems and preferred approaches can produce clearer goals for training and help avoid serious mistakes. In contrast to most medical therapeutic activity, in which the physician sees the patient over a longer time-span and has more opportunity to correct and adjust for error, the medical geneticist more often has only comparatively short contacts with counsellees. During these contacts, statements with lifelong consequences are often made without practical possibility for control in case of misunderstanding. Thus, medical geneticists have a strong need to keep up their proficiency, be aware of their shortcomings, and carry out their work in a setting with an optimal standard of quality control.

Secondly, although medical geneticists in many nations doubtless agree on some basic principles and practical approaches to moral problems, their views are in an 'oral tradition' rather than in normative guidelines. To leave the weighty questions of ethics raised in prenatal diagnosis to oral tradition alone ignores the need to clarify standards of practice and communicate to patients and policy makers what values and principles, as well as what specific approaches to problems, are most important to medical geneticists. In a field that has become in every way a medical specialty, medical geneticists are obliged to clarify their standards of practice. Such statements can assist in teaching younger colleagues and help the specialty present itself clearly to other specialities.

The third reason is that the future of medical genetics will be ethically more complex than that of the past. The potential for presymptomatic diagnosis of disorders of late onset, and for predictive screening of persons whose genotypes make them more susceptible to common diseases, will raise new issues of disclosure and public policy. Facing such issues with some help from guidelines is better than facing them simply with the strength of convictions. Preparations for the storms of the future can be made by consolidation of the wisdom of the recent past.

When the degree of consensus and variation between medical geneticists in 19 nations, mentioned earlier, has been studied and reported, the task will then be up to the professional societies in those nations to support the development of written statements affirming those principles and approaches which clearly have the most support. Whether real consensus can be achieved in international terms is an open question.

In my view, two reasons support a continuing interest in the questions of international consensus. First, many more persons of differing cultural backgrounds who are also at higher genetic risk will present themselves to medical geneticists in the future. Why? Mass migration of peoples and refugees will continue; borders between nations will also continue to 'soften' to encourage travellers, students and workers from other nations; the phenomenon of mass global communication will continue to reach into every culture to provide new information and education. Barring global biological or economic catastrophe, medical geneticists will increasingly see cases involving persons who do not share their cultural heritage but who are entitled to services. What weight should be given to cultural differences in ethics when faced with a real case? What kind of guidance do we have on the subject?

Secondly, the governments and peoples of our societies will not be content passively to allow the power to 'biopsy the human genome' to accumulate in the hands of a few without knowing more about the values and commitments of those to whom such power may be entrusted. Society may have been content in the recent past to allow the first increment of moral guidance in medical genetics to grow up naturalistically, as it were, out of the offices and clinics of medical geneticists. Members of society and its leaders realise that medical genetics and its allied fields are on the threshold of a new and more complex adventure in the diagnosis, treatment and prevention of genetic disorders. They will want to know the ethical commitments of those who propose to lead them across new moral ground well in advance of the journey.

ACKNOWLEDGEMENTS

The author gratefully acknowledges the help of Dr. Dorothy C. Wertz and Dr. Kare Berg in the preparation of this paper. The research for this paper was supported by a grant from the
Glenmede Trust Company, Philadelphia, PA.
[a] International Survey of Medical Geneticists. A study in progress supported by the Medical Trust, one of the Pew Memorial Trusts, administered by the Glenmede Trust Company, Philadelphia, PA. Report to be published in 1988 by Springer- Verlag, with the title "Ethics and Human Genetics: A Cross- Cultural Perspective."

ᵇ My views in ethics are based predominantly on premises described as "rule utilitarian." These views are described more fully in a chapter on ethical issues in prenatal diagnosis in Milunsky, A. (ed.): Genetic Disorders and the Fetus, 2nd edition. New York, Plenum Press, 1986, pp. 819-859.

ᶜ Early challenges to prenatal diagnosis on ethical grounds can be found in Kass, L.: Implications of prenatal diagnosis for the human right to life, in Hilton, B., Callahan, D., Harris, M., Condliffe, P., and Berkeley, B. (eds.)., Ethical Issues in Human Genetics: Genetic Counseling and the Use of Genetic Knowledge. Fogarty International Proceeding No. 13. New York; Plenum, 1973, pp. 185-199; Ramsey, P.: Screening: An ethicist's view, in Hilton, B. et al., above-mentioned volume, pp. 147-161; Lejeune, J.: On the nature of man. Amer. J. Hum Genet. 22, 121-128, 1970. Ethical arguments in favor prenatal diagnosis and selective abortion can be found in Hirschhorn, R.: Practical and ethical problems in human genetics. Birth Defects: Original Article Series 8, no.4, pp. 17-30, 1972; Milunsky, A: The Prenatal Diagnosis of Hereditary Disorders. Springfield, Ill., Charles C. Thomas, 1973; Fletcher, J. C.: The brink: The parent- child bond in the genetic revolution. Theological Studies, 33, 457-485, 1972.

BIBLIOGRAPHY

1 Antsaklis, A., Politis, J., Karagiannopoulos C., Kaskarelis, D., Karababa, P., Panourgias, J., Baussiov, M. and Loukopoulos, D.: 1984, *Selective survival of only the healthy fetus following prenatal diagnosis of thalassaemia major in binovular twin gestation.* Prenat. Diag., *4.*, 289.
2 Committee on Social Affairs, Parliament of Norway:1982–3, Report No. 91, pp. 11–12.
3 Dove, G. A. and Blow, C.: 1979, *Boy or girl: prenatal choice?* Brit. Med. J., *2*, 1399.
4 Fletcher, J. C., Berg, K. and Tranøy, K. E.: 1985, *Ethical aspects of medical genetics. A proposal for guidelines in genetic counseling, prenatal diagnosis and screening.* Clin. Genet., *27*, 199.
5 Fletcher, J. C., Hibbard, B., Miller, J. R., Rudd, N. and Shaw, M. W.: 1980, *Ethical, legal and social considerations of prenatal diagnosis.* Prenat. Diag. (Special Issue), Report of an International Workshop, Dec. 1980, 43–46.
6 Harris, M. (ed.): 1972, *Early Diagnosis of Human Genetic Defects,* Fogarty International Proceedings No. 6, Wash., D. C., U.S. Govt. Print. Ofc., HEW Publ. No. (NIH) 72–25, 143–145.
7 Kazazian, H. H.: 1980, *Prenatal diagnosis for sex choice. A medical view,* Hastings Center Rept., *10*, 17.
8 Lenzer, G.: 1980, *Gender ethics,* Hastings Center Rept., *10*, 18.
9 Littlefield, J. W.: 1972, *The pregnancy at risk for a genetic disorder,* N. Engl. J. Med., *282*, 627.
10 Milunsky, A.: 1977, *Know your Genes,* Houghton Mifflin, Boston.
11 Modell, B: 1985, *Chorionic villus sampling. Evaluating safety and efficacy.* Lancet, *1*, 737.
12 Modell, B.: 1986, *Some social implications of early fetal diagnosis,* In B. Brambati, G. Simoni, and S. Fabro, eds., Fetal Diagnosis During the First Trimester, Marcel Dekker, New York.
13 National Institute of Child Health and Human Development: 1973, Report of a Consensus Development Conference. Antenatal diagnosis. NIH Publication No. 79, 1–74.
14 Powledge, T. M. and Fletcher J. C.: 1979. *Guidelines for the ethical, social and legal issues in prenatal diagnosis,* N. Engl. J. Med., *300*, 168.
15 President's Commission for the Study of Ethical Problems in Medicine and Biomedical and Behavioral Research: 1983, Screening and Counseling for Genetic Conditions. Washington, D.C., U.S. Government Printing Office.

16 Rogers, E. M. and Showemaker, R. R., *The Communication of Innovations*; Rogers E. M., *Diffusion of Innovations,* In, Harris, M., ed.: 1972 Early Diagnosis of Human Genetic Defects, Fogarty International Proceedings No. 6. Washington, D.C., Superintendent of Documents, U.S. Government Print. Ofc., HEW Publ. No. (NIH) 72–25, 143–145.
17 Swiss Society of Medical Genetics: 1984, Medizinische Genetik, No. 12, p. 11.

U. WILKEN

THE NECESSITY FOR ETHICAL REFLECTION IN GENETIC COUNSELLING: ADVICE FOR PROFESSIONALS

THE ETHICAL AMBIVALENCE INVOLVED IN ASSESSING RISK FACTORS AND DIAGNOSIS

The objectives are:
- preconceptual genetic consultation before pregnancy
- prenatal counselling during pregnancy (following prenatal diagnosis)
- postnatal counselling after the birth of a handicapped child (viewed in terms of neonatal care and potential risk factors for future children).

These objectives should be guided by the premise that they are not directed against the disabled but rather seek to avoid disability, disease and retardation. To this extent such counselling is preventive. However, the practice of prenatal diagnosis not only leads to the diagnosis of disability, disease or imminent disorders, and to any feasible treatment, but it also frequently leads to a termination of the pregnancy. This is due to the limited possibilities of prenatal therapy in general. Here the prevention of imminent disabilities becomes the prevention of the birth of a disabled child. Abortions should, therefore, not be understood in a traditional medical sense as preventive measures applied in the interests of the child's well-being; nor should they be euphemistically described as therapeutic abortion.[4] No matter what the justification and indications might be, one should realise that prenatal fetocide is the result and not the therapy. The only possible therapeutic effect might be on an expectant mother where the continuation of the pregnancy could have potentially negative physical or psychological consequences. Or, abortion could perhaps also be viewed as justifiable euthanasia for the developing child if it, as a newborn, cannot be kept alive with modern medical technology, or if he or she will suffer such severe damage as could be compared with the cessation of life-support systems of persons who are considered brain dead.[7]

The question then arises as to whether early intrauterine selection and elimination of defective life is morally more acceptable than postnatal euthanasia of a disabled child immediately after birth. This question becomes all the more pertinent as abortion may be requested not just as a result of a prenatal diagnosis of embryo or fetal damage, but also when disorders or disabilities can be predicted with only a very low probability. This is the case with disabilities resulting from some genetic disorders or from various infectious, physical, chemical, mechanical, or immunological hazards during pregnancy. In such circumstances, the

73

E.K. Hicks and J.M. Berg (eds.), The Genetics of Mental Retardation, 73–77
© *1988 by Kluwer Academic Publishers.*

possible intrauterine fetocide of healthy children is accepted in order to minimize the risk of giving birth to a potentialy disabled child.[7]

A further conflict may arise when prostaglandin-induced abortion is done late in pregnancy ($\pm$ week 22), when the fetus, if left undisturbed, would potentially have a level of development enabling it to be born alive, but due to its prematurity cannot be kept alive. At this stage in fetal development the birthweight ($\pm$ 600 grams) falls short by a mere 100-300 grams (or 2-3 developmental weeks), when intensive pediatric support would commonly be applied in the event of premature birth. "The preservation of life of premature, underdeveloped, disabled children, including those children possibly disabled as a result of intensive medical treatment, very closely approaches the zone where abortion is performed based upon diagnosis that the child may be disabled."[2]

We must consider what repercussions this seemingly ambivalent attitude between infanticide and intensive medical intervention has on professional practice and ethics. We must also consider how it can be ethically justified that active euthanasia is legally possible via intrauterine abortion up to the 22nd week, but euthanasia is not allowed in the case of children born disabled, the terminally ill, and the elderly, of whom it is presumed that they are a burden to themselves and to others.[2] In this respect the problem is not confined to mother and child but encompasses society in general, especially those who are seriously ill, the elderly, and the disabled.

Social consciousness currently implies an attitude of 'do what is do-able.' Society is no longer willing to accept premature limitations to practicability because of 'fate' – however defined. Consequently, it has become necessary to reflect upon the 'new ethos', asserted by those involved in the choice of whether or not to abort a potentially retarded child. This reflection should be rational and ideologically unbiased. With regard to the lifting of taboos against developing life, as well as those against sanctity of life, which is, from a contemporary medical viewpoint, sometimes called into question, it is necessary to establish a responsible and common consent which will help to confine arbitrariness, abuse and presumption. Further reduction of conflicts and the development of ethical consensus will result from earlier prenatal diagnostic possibilities in the first trimester with chorionic biopsy. This should not, however, distract from research into prenatal and neonatal therapy. Although the early embryonic stage of biological life should not be refused the designation 'human', during this early stage there are fewer disturbing consequences, and, in terms of ethics, more humane provisions for the tolerance of the dilemma involved with termination of pregnancy. Developments permitting early prenatal diagnosis and hence earlier termination of pregnancy will generally result in minimizing the danger of crossing over bounds in the above-mentioned fundamental ethical concerns.

ETHICAL CONFLICTS DURING GENETIC COUNSELLING

The ethical ambivalence which arises during counselling may conflict with its goals, i.e. either principally to protect life, even if it is disabled, or, in the case of imminent diabilities, to prevent such life since it is unreasonable to place such demands on the mother. This ethical ambivalence must be endured and a decision must be reached. From the pragmatic point of view that an abortion generally requires professional assistance, it follows that the problem is not only that of one individual, but involves the doctor as well as the expectant mother. While a binding ethical consensus does not yet exist, the person conducting the counselling has decisive importance.

Because ethical reorientation and the legal definition of feasible criteria for abortion are asynchronous to technological and scientific progress, the ethical and legal framework of definitions and criteria must still be developed. Within existing general legal guidelines, conflicts can occur with regard to minimizing conflict of interests in individual cases. During genetic counselling, the counsellor's personal attitude may dominate and thereby hinder an autonomous decision by the expectant mother. Conflicts can also arise when the expectant mother is unwilling to assume responsible parenthood, but prefers to follow her inclinations or pleasure (which may be difficult to accept socially or ethically), making the physician, counsellor, and also the child into agents in the obtainment of her individual happiness. Even if the doctor is legally responsible to provide a credible medical opinion, the counselling goal is to help the patient in a free decision. There should be no projection of the counsellor's personal ethical position.[6]

In order to minimize an incongruous leeway in the decision- making process, a normalising catalogue of criteria could provide an objective aid for counsellor and patient. However, this may lead to an over-schematisation in solving conflict of interests in relation to the individual's capacity to cope with stress and suffering. At the same time, a certain consensus is required beyond which abortion on the basis of prenatal diagnosis is not acceptable. These limits could be set by exclusion of criteria such as, for example, selection of gender, or disorders carrying a risk not exceeding the life-time risk which everyone has (e.g., in the case of psoriasis). With conditions such as anencephaly or, in contrast, psoriasis, possible standards for or against legal termination of pregnancy are relatively easy to develop. The current counselling situation, however, is especially problematic with regard to disorders such as Down's syndrome, dysmelia, spina bifida, phenylketonuria, and X-linked hereditary disorders (e.g. muscular dystrophy, hemophilia).

Careful consideration of potentially treatable disorders must be made prior to any decision to terminate pregnancy, particularly when such potential treatment could enable an affected person to lead a relatively independent life within the community. Ambivalence in this regard is sometimes experienced by mothers who are raising Down's syndrome children, and who have come to accept and love these children.[10] On learning, through prenatal diagnosis in a subsequent pregnancy, that the current fetus is also affected, even these mothers generally express a desire to terminate the pregnancy.[5]

Many counselling centres, motivated by a general ethical consideration for life and the right to life, cooperate with organisations for handicapped people, using their experiences to help in the decision-making process regarding abortion.[1] It must, of course, be recognised that mothers of disabled children usually do not want to give birth to another disabled child. Similarly, the staff of institutions for the handicapped, though devoted to the therapy, care, and education of affected children, rarely hesitate to support a parent's decision for an abortion in the event of a subsequent affected pregnancy; nor would they hesitate to abort if they were themselves personally confronted with this situation.[3] However, the following general statement could apply: "If anyone is permitted to make a decision in such a case, it should be those who must bear the burden of the disabled. The goal of such counselling is not to leave these people alone with their decision, but to help them reach an autonomous decision without sparing them the responsiblity of it.[1]" An ethical consultation should be based upon the freedom to decide for or against abortion, and this decision should be the responsibility of the expectant mother in so far as this is reasonably possible.

Another consideration in a decision about abortion may be that society often does not offer adequate psycho-social and economic help for the education and life-time welfare of disabled children. As long as disabilities are viewed as constituting individual fate which must be individually endured, and not as a common societal obligation, the attitude towards the birth of a disabled child will remain negative, and will lead to social stigma.[8,9] Moreover, from fear of social discrimination, blame and stigmatisation, abortion, no matter at what price, will be regarded as the only possible solution. Accordingly, one may ask if abortion after prenatal diagnosis, representing the pragmatic 'new ethos', is socially and ethically acceptable. It might also lead to misuse of ethics as a cover-up for antisocial and political goals, legitimized at the cost of afflicted persons.

Because social practice and the formation of ethical consciousness occur within a dialectical process, the afore- mentioned conflicts and their origins must be analysed and considered during consultation with the expectant mother. If, during the counselling sessions, nothing more than *minima moralia* becomes apparent, one should not speak too hastily of a 'selling out' of human values in the face of a pro abortion decision. We should be aware that this ethical problem is largely caused by the lack of solidarity with the weak, the old, the ill, and those on the margins of our society. To achieve a greater understanding, tolerance and valorisation of the disabled, a general re-orientation of the social and family structures is necessary in our society, a society that is still seeking humanitarianism.

BIBLIOGRAPHY

1 Eibach, U.: 1985, *Konflikte in der humangenetischen Beratung,* Diakonie – Zeitschrift des Diakonischen Werkes der Evangelischen Kirche in Deutschland, *11*, 110.

2 Hepp, H.: 1982, *Schwangerschaftsabbruch aus kindlicher Indikation – anthropologisch-philosophische Aspekte des Arzt-Patienten-Konfliktes,* In, P. Boland, H. A. Krone, and R. A. Pfeiffer, eds., Bamberger Symposion: Kindliche Indikation zum Schwangerschaftsabbruch, Milupa AG, Friedrichsdorf/Taunus, 33–48.

3 Lenz, W.: 1982, *Die sogenannte genetische Indikation zum Schwangerschaftsabbruch,* In, P. Boland, H. A. Krone, and R. A. Pfeiffer, eds., Bamberger Symposion: Kindliche Indikation zum Schwangerschaftsabbruch, Milupa AG, Friedrichsdorf/Taunus, 11–15.

4 Pfeiffer, R. A.: 1982, *Schlusswort,* In, P. Boland, H. A. Krone, and R. A. Pfeiffer, eds., Bamberger Symposion: Kindliche Indikation zum Schwangerschafts-abbruch, Milupa AG, Friedrichsdorf/Taunus, 160–170.

5 Recke, von der, S.: 1985, *Verkraften wir noch ein behindertes Kind?* Zeitschrift Zusammen: behinderte und nicht behinderte Menschen, Freidrich Verlag Velber, *5*, 16.

6 Schroeder-Kurth, T.: 1982, *Schwangerschaftsabbruch- Ethische Probleme bei der genetischen Beratung,* Geistige Behinderung, *21*, 224.

7 Tettenborn, U.: 1985, *Schwangerschaftsabbruch aus genetischer und ärztlicher Sicht,* In, J. Reiter, and U. Theile, eds., Genetik und Moral, Beiträge zu einer Ethik des Ungeborenen, Grünewald Verlag, Mainz, 109–114.

8 Wilken, U.: 1984, *Grundannahmen einer pädagogischen Theorie offensiver Rehabilitationspraxis – sechs Thesen,* Behindertenpädagogik, Vierteljahresschrift für Behindertenpädagogik in Praxis, Forschung und Lehre, *23*, 226.

9 Wilken, U. 1984, *Zur Geschichte von Einstellungs-determinanten gegenüber Körperbehinderten,* Die Rehabilitation-Zeitschrift für alle Fragen der medizinischen, schulisch-beruflichen und sozialen Eingliederung, *23*, 81.

10 Wilken, E.: 1985, *Sprachförderung bei Kindern mit Down- Syndrom,* 4th ed. Marhold-Verlag, Berlin.

DISCUSSION: PRENATAL DIAGNOSIS

Screening:

A distinction was considered necessary between screening of the general population, or groups within it, and diagnostic undertakings on specific individuals, and that this distinction must be made clear to the public. Moreover, what groups should be screened (e.g. among women planning pregnancy and among newborns), and where, how, and when, require careful consideration.

The general expansion of testing capability creates a potential problem from the laboratory point of view. Specifically, the increased laboratory workload associated with this expanded capacity for chromosomal or other testing raises the risk of error, unless there is a concomitant growth of appropriate and reliable laboratory facilities.

Gender selection:

The capability to make an early diagnosis of the gender of a fetus constitutes a growing problem in prenatal diagnosis. In societies where male children are preferred, it is not unusual for families to request termination of pregnancy if the fetus is female. It was agreed that this is a socio-cultural problem rather than a biomedical one, since gender is not an abnormality.

Professional moral consensus and parental autonomy:

Does a moral consensus already exist among genetic counsellors and, if so, what procedures should be followed to qualify its nature and its parameters in a pluralistic professional, juridical, and cultural world?; if not, does it behoove professionals to try to achieve such a consensus? If the choice is to do so, what about the procedural difficulties? Moreover, what would be the professional and social consequences of such a consensus? Would it, in effect, amount to a professional 'hit-list', since such a consensus would perforce involve a view as to what abnormalities or deformities must or should be present in the fetus in order to warrant termination of pregnancy? To what degree would such a consensus affect parental autonomy in the decision-making process about the future of a fetus or newborn?

With regard to the last mentioned point, the degree to which parental autonomy in such a decision-making process can be exercised is becoming increasingly problematic, given the complexity of the technological data which must be comprehended in order to exercise effective decision making.

79

E.K. Hicks and J.M. Berg (eds.), The Genetics of Mental Retardation, 79–80
© 1988 *by Kluwer Academic Publishers.*

Other factors which compound the difficulty of parental autonomy in deciding the future of a fetus or newborn include:

a. the apprehension of those counselled to accept decision- making responsibility;
b. the presence or absence of medical/ethical opposition – e.g., where the population is simultaneously Catholic and highly educated, decision-making can be a problem when pregnancy termination may be advisable;
c. parental attitudes to aborting a fetus. Many parents of a terminated pregnancy express the desire to 'know the fetus as a part of the family,' either through a religious ceremony of some type after termination, or by means of photographs of the abortus;
d. the advisability of a period of mourning prior to initiating a subsequent pregnancy;
e. the maintenance of professional integrity and communication between counsellor and those counselled;
f. the degree of certainty of the diagnosis.

When, and to whom, should prenatal diagnosis be available:

Should prenatal diagnosis be available on demand, or only when particular indications are present to warrant it? Moreover, at what maternal age is prenatal diagnosis indicated, and does parental anxiety *per se* constitute an indication for prenatal diagnosis?

When considering these issues, the positive side of prenatal diagnosis should be kept in mind. Although it is usually undertaken in cases where there is a risk of fetal abnormality, prenatal diagnosis is very often life-preserving, i.e. test results are usually normal. This eliminates the anxiety caused by risk of fetal abnormality, with a consequent decision to continue with a pregnancy which might otherwise have been terminated.

SECTION II

POSTNATAL DIAGNOSIS

SCREENING FOR HEMOGLOBINOPATHIES IN NEW YORK CITY

In developed countries the newborn period is a convenient time for genetic screening, because it is a time when almost every infant and his or her mother enters the health care delivery system. The detection of babies with a particular genotype may allow intervention which can prevent the development of symptoms, as in some metabolic diseases, or more adequate treatment for the early symptoms of a disease process which as yet cannot be prevented, as in sickle cell disease. Newborn screening programs require tests simple and accurate enough to be applied on a large scale; the efficacy of screening further depends upon a means for quick and efficient follow-up of affected individuals.

In most cases, detection of heterozygous carriers for recessive genetic disease cannot be accomplished by the same simple tests used to detect the affected homozygotes, so that the problem of transmitting this information to the much larger number of carriers does not have to be faced. An exception is found among the hemoglobinopathies where the hemoglobin electrophoresis used to detect homozygotes for S,C and other hemoglobin variants also detects heterozygotes. In areas of high incidence, screening of women in prenatal clinics for hemoglobinopathies is also practical.

This paper describes our experience with screening for sickle disease and sickle trait in a region of New York City where the frequency of gene carriers is high. Although the hemoglobinopathies do not themselves lead to mental retardation, the subject is appropriate to this symposium because it presents a model of what may become a much commoner situation in genetic surveillance, as more and more diseases become amenable to such carrier detection and prenatal diagnosis.

Sickle cell anemia and related hemoglobinopathies, such as those involving Hemoglobin C or the combination S/thalassemia heterozygote (which will be referred to collectively as "disease hemoglobinopathy"), have several properties which are important in considering the impact of carrier detection:

(1) They occur almost entirely in particular racial groups, which, at least in the United States, are often overrepresented among the poor and socially disadvantaged
(2) They are very variable in their impact upon the health of affected individuals
(3) There is some dispute as to the effects on the health of heterozygotes
(4) Prenatal diagnosis is available.

E.K. Hicks and J.M. Berg (eds.), The Genetics of Mental Retardation, 83–90
© 1988 *by Kluwer Academic Publishers.*

HISTORY OF THE SICKLE CELL SCREENING PROGRAM

In the United States a federally funded program called the National Sickle Cell Control Act was established in 1972, which provided money for research, education, screening and counselling. In the 1970s several states set up mandatory carrier testing programs of the black population, usually without availability of appropriate testing methods, follow-up counselling, or education of the community. There was often distress and confusion among those screened, and discriminatory practices against carriers, as a result of ignorance and/or racial prejudice. Widespread outcries against the screening program were heard, particularly from the black community, and the principle that screening should be voluntary was widely endorsed. In 1978 funding for sickle cell disease was incorporated into a broader federal program on genetic disease. Genetic programs are now generally state or regional in nature.

In New York State mandatory screening of the newborn for sickle cell disease began in February 1975, when this disease was added to the existing phenylketonuria screening program along with 6 additional inherited metabolic disorders. This program was initially ineffective, since no money was allocated for follow-up of presumptive positives or specimens needing retesting.[2] Since 1979 better follow-up programs have been in place[1], with effective follow-up of positive infants increasing to close to 100% in 1985.

SICKLE CELL DISEASE: FREQUENCY AND FOLLOW-UP OF NEWBORN SCREENING

Statewide there are 10 times more positive diagnoses for the hemoglobinopathies than for all the other 7 metabolic errors combined. Almost all affected individuals are found among Black or Hispanic births. These comprise about 25% of births in New York State, and 50% of births in New York City. The proportion varies greatly from region to region, and within New York City, from one hospital to another. In the 3 hospitals in Upper Manhatten described in this paper, 70% of a total of approximately 12,000 births per year are to Black and Hispanic women. Of all infants with sickle cell disease in New York State, 83% are born in New York City.

The electrophoretic test used detects not only SS but also SC, SD and S-β thalassemia. These variants are relatively common because of the high frequency of the C allele and thalassemia genes in the population at risk for SS disease. In some parts of West Africa, from which the ancestors of most Afro-Americans came, the C allele is as common or more common than the S allele. Local populations, differing greatly in the relative frequencies of the C and S allele, have become merged in America. The relatively high sickle gene frequency among the Hispanic population, which comes largely from the Carribean Islands, Central and South America, is a result of the racial admixture which took place between the various racial groups (Negro slaves, native Indians, Europeans, Orientals) upon coming to America, and which still continues in areas such as New York City.

In 1984 the frequency of sickle cell disease was 0.3%, or 1/300, among births classified as Black or Hispanic in our 3 Manhattan hospitals. Of the 25 affected babies, 16 had SS disease, 4 had SC disease, 1 each had CC and S/thal disease and 3 had other variants.

The results of the newborn screening are returned from the state laboratory to the hospital of birth, which then has the resposibility for follow-up. About 2% of tests must be repeated, owing to inadequate samples or problems such as transfusion. An attempt is made to notify the parents of all babies suspected of having sickle disease (by phone or by mail), further testing is provided, and those identified as diseased are referred for comprehensive care to special sickle cell clinics. Genetic counselling is provided both at the time of retesting, and after the results are obtained. In 1985 there were a total of 209 births initially identified as having sickle disease in New York City (Thomas Carter, personal communication). All were confirmed on further testing, although in some cases the specific hemoglobinopathy needed further clarification.

The early identification of the disease is expected to decrease early morbidity, which is often the result of increased susceptibility to infection in the affected infants. About 15% of cases currently die before 2 years of age. A survey of infant deaths due to sickle cell disease, before screening follow-up was effective, showed that in the majority of cases the parents were unaware of the presence of the disease before the life-threatening episode. Aggressive treatment of febrile illness, prophylactic antibiotics and immunization to pneumococcus after infancy appear to reduce mortality.[5] The effectiveness of this is being evaluated in a population followed from birth.

CARRIER DETECTION IN SCREENING PROGRAMS: FREQUENCY AND IDENTIFICATION OF COUPLES AT RISK FOR CHILDREN WITH DISEASE

The ability to detect heterozygotes in the newborn screening program provides a means of identifying couples at risk for a child with a sickle cell disease, because at least 1 parent in such cases must be a carrier. Since 1980 our 3 Manhattan hospitals have been attempting to contact the parents of all babies identified as heterozygotes, explaining the results of the test and requesting both parents to come in for testing and counselling. With adequate staffing, up to 90% of couples with a baby with a sickle cell trait can be contacted. Data have been analyzed for 1 hospital, with approximately 4,000 births per year, for the years 1984 and 1985. Over this time period there were 403 babies with sickle cell trait identified in newborn screening: 367 mothers and 209 fathers were tested for trait status. In some cases mothers were already aware of having the trait because of screening in the prenatal clinic. In 47 cases the father was unavailable for testing even when the mother was found to be a carrier. During the 2-year period, 12 couples at risk for sickle cell disease were identified via the testing of the parents

86

TABLE 1

Newborn and prenatal clinic detection of sickle cell disease and trait
Presbyterian Hospital in the City of New York, 1984 and 1985

	Newborn Screening*		Prenatal Clinic**	
Number tested	8046		4350	
Number with disease:	12		15	
hemoglobinopathy SS		9		4
SC		3		5
CC		0		1
S or C thal		0		5
Number with trait	403		454	
Number counselled	367		330	
Number of fathers	403		454	
untested with	367		330	
carrier mother	47		201	
Number of couples at risk detected	24***		24	

* 64% Black and Hispanic
** 90% Black and Hispanic
*** Includes 9 couples also identified through prenatal screening

of newborn heterozygotes. Twelve additional couples were counselled following the birth of an affected child (Table 1).

For an 18-month period parents were classified according to ethnic group. Two main Hispanic groups are identifiable in this hospital's catchment area: those from the Dominican Republic and those from Puerto Rico. We classified as Black all those whose race was given as black but who were non-Hispanic. Table 2 shows the frequency of heterozygotes for the S and C allele in these 3 groups. The frequency of carriers of the S gene among Blacks is about 8%, agreeing with other figures from United States black populations. The frequency among both those from the Dominican Republic and Puerto Rico is also high, 3–4%. In the Hispanics the frequency of carriers detected agrees well with that expected by calculating the gene frequency from the frequency of homozygotes, and assuming Hardy-Weinberg equilibrium. However, in those classified as Black, non-Hispanic, there is a deficiency of heterozygotes compared to that expected on the basis of Hardy-Weinberg. This is almost certainly due to the obvious heterogeneity of this classification, which includes people from the West Indies such as Jamaica, new West African immigrants, and descendants of American slaves.

In some areas of New York State screening for hemoglobinopathies is also carried out on all pregnant women in prenatal clinics, on the grounds that detection of women with previously undetected sickle cell disease aids in the

TABLE 2

Frequency of sickle cell disease and trait in Black and Hispanic births (18 months, Presbyterian Hospital in the City of New York, 1984–85)

	Parents Dominican	Parents Puerto Rican	Parents Black, non-Hispanic
Number of births	2900	688	1344
Number with disease SS	1	0	5
SC	1	0	1
Number of carriers AS	108	23	106
AC	25	7	29
Gene frequency, $q = \sqrt{\text{frequency SS}}$	.017	–	.061
Expected frequency of AS = 2pq	.034	–	.115
Observed frequency of AS	.037	.033	.079

management of their pregnancies. However, counselling and testing of mates is also offered whenever an affected or carrier woman is detected. The patients in the public prenatal clinics in the 3 Manhattan hospitals being discussed are almost entirely Black or Hispanic. In the years 1984 and 1985, at the same hospital described above, 0.3% (15/4350) of women were found to have disease; this is identical to the frequency in Black and Hispanic newborns. Many of the women with disease were unaware of their diagnosis or its genetic implications. About 10% of women (454/4350) were found to be heterozyous for S,C or a rarer allele. In all those women with disease or trait an attempt was made to bring in the father of the pregnancy for carrier testing. This was successful in only 54% of cases; resulting follow-up of mates identified 24 couples at risk for a child with a sickle cell disorder, who were then given further counselling, including information on the availability of prenatal diagnosis.

There is obviously overlap between the screening initiated through newborn testing and that initiated through screening of prenatal clinics. However, each type of screening complements the other. Some couples at risk will not have babies with trait or disease, and will therefore go undetected in the newborn screening program. On the other hand, some babies with trait or disease are born to mothers who have not had prenatal care in the hospital clinic, and more fathers are willing to come after the baby is born rather than during a pregnancy. Newborn screening also identifies cases where the father is a carrier but the mother is not.

In the 2 years under discussion, there were a total of 39 heterozygous couples at risk for sickle cell disease detected via screening. Of the 24 couples identified via prenatal screening, there were 11 who had AA babies or spontaneous abortions, and would therefore not have been detected via newborn screening. There were 13 couples who were detected via newborn screening who were not

88

identified via prenatal screening. In 6 cases where babies had disease or trait the mother had been identified as a carrier but the father had not been available for testing prenatally. Nine newborns with trait or disease had mothers who did not receive their prenatal care in the same hospital. Only 3 of these were aware of their carrier status through prenatal screening in another hospital.

Of the 39 at-risk couples identified over the 2 year period, only 3 had previous knowledge of their risk, all because of a previous affected child. Screening allowed 26 couples to know of their risk other than through the birth of an affected child.

Optimally, screening programs would attempt to identify parents who were B-thal carriers also, since they can produce children with sickle-thal disease when mated to a sickle cell carrier. This requires performing an MCV as well as hemoglobin electrophoresis on all individuals, with follow-up of those showing a low value. In our program this is done during the prenatal screening, but not as yet during newborn follow-up.

PRENATAL DIAGNOSIS

Couples at risk for a child with a sickle-cell disorder are counselled about the nature of the disease and the availability of prenatal diagnosis. Current techniques which make the diagnosis directly from DNA analysis of the mutant gene allow prenatal diagnosis even when the father's genotype is not known. The risk of sickle cell disease when the mother is a known carrier is equal to 1/4 of the heterozygote frequency (the sum of both the S and C heterozygotes) in the population. For our Black population this is about 1/40, and for the Hispanic groups about 1/100. Few such pregnancies undergo prenatal diagnosis, both because the risk is not perceived as high, and because of the limited capacity of laboratories doing the prenatal diagnosis. In the case of a mother with sickle cell disease and a father of unknown status the risk is twice as high, and some such pregnancies have been monitored. One cannot rule out a possible S/thal or S/C heterozygote when the father's status is unknown.

Of the 24 at-risk couples detected during prenatal care in 1984 and 1985, there were 8 couples who were detected too late in pregnancy for prenatal diagnosis to be carried out, 5 who elected not to have the test and 11 who had prenatal diagnosis via amniocentesis and DNA analysis. There were 4 affected fetuses detected, 2 with SS and 2 with SC disease. The 2 couples with an SS fetus chose to terminate the pregnancy.

PROBLEMS ILLUSTRATED BY THE SICKLE CELL SCREENING PROGRAM

1. The importance of screening for sickle cell disease has not been universally accepted by the population at greatest risk. As one black health professional[4] said,

"Sickle cell anemia appears to have become a priority among biomedical professionals, whereas it is not a primary concern of the affected population. One wonders why this has become such an important issue, in contrast with the socioeconomic disadvantages faced by African-Americans which are clearly associated with poor health status."

Most of the population at risk, particularly the Hispanic population, are unaware of the sickle cell disorders, and of their risk, and may perceive screening as motivated by racial discrimination rather than a concern for their health. However, the frequency and severity of the disease, and the ability to intervene in meaningful ways, do appear to merit screening of the population at high risk, just as is done for much rarer metabolic disorders such as phenylketonuria, of which the screened population is also largely ignorant.

2. The screening system appears to be inefficient because all individuals in the population are screened, whereas almost all the risk lies in a sub-population identifiable by ethnic status. However, since the same collection system is used for all the diseases screened for in the newborn period, the effort involved in restricting screening for particular diseases would probably not be cost effective.

3. The aggressive follow-up of heterozygotes detected in the newborn and pre-natal screening programs, which are obstensibly carried out to detect disease, not trait, can be criticised on the grounds of violation of privacy concerning the results of a test for which informed consent has not been given. However, there is also the question of medical responsibility to communicate the results once they have become available, and possible liability should an affected child be born because of lack of knowledge of the test results. We have chosen to act on the information available, while realising that effective education and voluntary screening would be preferable. The reaction of the population seen in individual counselling sessions has been positive: they have accepted it as part of their routine medical care, which commonly involves testing whose purpose is not explained beforehand. However, a program like this requires a great deal of effort for follow-up, and skilled and dedicated personnel for counselling. It should not be embarked upon unless there is a serious commitment to spend the time necessary to provide effective communication.

4. There are great difficulties in obtaining follow-up, particularly of fathers, in a population which often lives in poverty, is extremely mobile, has inadequate access to health care, and has a very high rate of out of wedlock births and unstable unions. This is basically a social problem, which applies to many aspects of health care for the disadvantaged, but the need for rapid follow-up of new-borns, and for the testing of both parents when the mother is found to be a carrier, makes it especially critical in this situation. It is of interest that testing of fathers is more readily achieved after the birth of a child with trait than during the prenatal period when only the mother's carrier status is known.

5. The variable expression of the disease, which may cause infant mortality,

severe or mild morbidity in childhood, or may remain relatively symptomless, makes it difficult to explain to those at risk. Perception of the severity of the problem, and response to a positive prenatal diagnosis, may depend upon whether the couple have had experience with a severely affected child or other relative, a mildly affected individual, or no personal experience with the disease. When a couple with no first-hand knowledge of the disease is considering prenatal diagnosis we give them the opportunity to meet patients and parents in the clinic which provides comprehensive care for patients with disease. Our experience over the last 5 years with prenatal diagnosis has been that fewer than half of couples choose to terminate an affected pregnancy, even though they elect prenatal diagnosis.

6. The possible distress caused by detection of carrier status in a healthy individual may lead to lowered self image, and discrimination in hiring, or in insurance practices. The distinction between having sickle cell "trait" and sickle cell "disease" is often not well understood by the patient, other health professionals or the community. Individuals known to have trait have been excluded from a wide variety of jobs, often as a result of ignorance, but sometimes as an excuse for racial discrimination.[3] Education of the community at risk, and of society at large, is lacking. A large proportion of those parents who are contacted through newborn screening programs are unaware that their offspring are being screened for genetic disease. The likelihood for confusion is particularly increased when the child has the trait only, and yet the parents are told to come in for counselling and testing. In these cases it is essential to stress the fact that the baby is not sick.

7. There is a completely unsolved problem concerning what should be done to preserve the information about carrier status which is obtained at the time of newborn screening. About 7,000 carriers of S, C or other rarer alleles are born in New York State every year. Since we have no permanent health records in the United States, this information will usually be lost to the individual when he or she becomes old enough to use it to make reproductive decisions, or to avoid situations which could lead to problems because of the carrier status. Any suggestion to make the results of genetic testing a part of some permanent record, such as a health or insurance card, may be met with objections to having a label which implies genetic abnormality or inferiority. After all, we all are heterozygous for some deleterious genes, but most of us are allowed to remain blissfully ignorant of what they are.

BIBLIOGRAPHY

1 Grover, R., Shahidi, S., Fisher, B., Goldberg, D. and Wethers, D.: 1983, *Current sickle cell screening program for new-borns in New York City, 1979–1980,* Amer. J. Pub. Hlth., *73,* 249.

2 Grover, R., Wethers, D., Shahidi, S., Gross, M., Goldberg, D. and Davidow, B.: 1978, *Evaluation of the expanded new born screening program in New York City,* Pediatrics, *61,* 740.

3 Holden, C.: 1981, *Air Force challenged on sickle cell trait policy,* Science, *211,* 257.

4 Randolph, W.:1983, *Social issues in perinatal screening for sickle cell anemia,* Urban Health, August, 32.

5 Warren, N. Carter, T., Humbert, J. and Rowley, P.: 1982, *Newborn screening for hemoglobinogathies in New York State: experience of physicians and parents of affected children,* J. Pediat., *100,* 373.

M. BARAITSER

THE USE OF THE COMPUTER IN THE DIAGNOSIS OF DYSMORPHIC SYNDROMES

The accurate diagnosis of a specific multiple malformation syndrome is of considerable importance to those who need to know the chances of a similar event happening in a subsequent pregnancy. About 7 per 1,000 children are born with more than 1 malformation and, whereas chromosomal abnormalities will account for about a half of these, the rest need particular scrutiny in order to differentiate between those conditions with a high recurrence risk and those which might not be genetic at all.

The problem for the clinician is that there are more than 1,000 recognisable non-chromosomal malformation syndromes and though some are rare, the syndromes make, as a group, a large impact on paediatric practice in the industrialised world. One in every 250 will have multiple malformations and normal chromosomes, and the vast majority of parents will want to know the diagnosis and discuss any events that happened during pregnancy or at birth that might be related to the diagnosis.

APPROACH

In the past, clinicians attemped to make a diagnosis by comparing the clinical picture with cases seen in their own experience, by consulting colleagues and by looking for similar cases in the literature. The exercise has become progressively more difficult as more and more syndromes are delineated and described.

Smith's monograph[4] contains a useful appendix of lists of syndromes manifesting particular abnormalities. Unfortunately, only syndromes covered in the book are listed and these make up, at most, 20–30% of those reported in the world literature. Finding recently reported malformation syndromes that are not in the textbooks is difficult as they are not referenced in Index Medicus under specific malformations. It is also likely that many malformation syndromes have not yet been described in the medical literature, as only individual cases have been seen by different clinicians.

Faced with a child or infant with multiple malformations there are 2 possible pathways to reach a diagnosis:

I. Gestalt recognition

Some syndromes are so distinct that recognition is immediate without a need to detail and evaluate the individual dysmorphic features.[2] Immediate recognition occurs in 2 ways:

1. Identification is possible because the features are grossly abnormal and therefore distinct. For instance, the combination of syndactyly involving all the fingers

E.K. Hicks and J.M. Berg (eds.), The Genetics of Mental Retardation, 91–97
© 1988 by Kluwer Academic Publishers.

and toes leaving the thumbs and big toes free, in conjunction with a tall skull shape, beaked nose and proptosed eyes, is so distinct that an immediate diagnosis of Apert's syndrome is nearly always possible – provided the clinician has seen the condition before.

2. Recognition is possible because the dysmorphic features, although minimal and in themselves not differing markedly from normal, form a picture which is distinct. The picture is clearcut because the brain is an excellent computer and can put together small variations to form a recognisable whole. How well it performs is clear when 10 children with Angelman's syndrome (Happy Puppet) are placed alongside and the similarities in the facial features become evident, even though the facial dysmorphism is slight and a pictorial distinction is difficult to give in words. The same applies to fragile – X mental retardation. This used to be characterised as non-specific or non-dysmorphic and even as 'pure' mental retardation in the apparent absence of relevant dysmorphic features. Now that the condition can be diagnosed by chromosomal analysis it has become clear that affected males are dysmorphic (they have big ears, a long face and a prominent jaw), enabling clinical recognition or suspicion by Gestalt in a large proportion of cases.

Among other examples, the FG syndrome is a diagnosis suggested in males (it is X-linked) by the presence of a relatively large cranium with broad forehead, mental retardation, severe constipation and minor skeletal changes. Diagnosis is difficult but as the number of published cases is increasing the minor dysmorphic features are beginning to form a pattern.

It should also be noted that Gestalt recognition in this group might be age dependent. For instance, in Williams' syndrome, it is difficult to make a diagnosis in the neonatal period especially in the absence of hypercalcaemia and a supra-valvular aortic stenosis. However, with passage of time, the facial features (which include sagging cheeks, supra-orbital fullness, minimal hypertelorism, thick lower lips, a long philtrum, and a stellate iris pattern) become clearer and the diagnosis is easier to make by Gestalt.

II. Attempted recognition by using diagnostic handles

If Gestalt recognition is not possible, the diagnostic procedure becomes more difficult. The clinical impression of a syndrome without immediate identification is usually followed by endless searches in books for similar cases. The success of this approach depends upon the presence of good handles. A handle is a dysmorphic feature which is definitely abnormal, i.e. it does not merge into the known variation that occurs in the population. Syndactyly of toes 2 and 3 is common and therefore a poor handle. A good handle would be syndactyly of fingers 4 and 5 or a malformed ear or anal atresia.[3] These dysmorphic features, in combination with others, might lead to a specific diagnosis, if there were lists available where the appropriate combinations could be consulted. The other reason why a handle might be a poor vehicle of entry into the literature is because it occurs in so many malformation syndromes. For example, low set ears have a poor discriminating value in deciding between the various possible syndromes.

At a more practical level, it would be of little use to ask the computer for all the syndromes with low set ears because the list would be too large to be manageable.

Before embarking on the search, the clinician must differentiate between a malformation, a disruption and a deformation. The vast majority of the syndromes under discussion are malformation syndromes. A disruption is characterised by extrinsic disturbances which alter the morphology of a potentially normal field of development. Similarly, a deformation results from an external force which alters the shape of otherwise intrinsically normal tissue. Many extrinsic causes will be evident from the history; for instance, maternal diabetes, fetal alcohol syndrome and fetal hydantoin syndrome all fall into this category. Amniotic bands causing multiple malformations and the effects of fetal compression, for instance femoral hypoplasia – unusual face syndrome, are other examples. A good family, pregnancy and birth history is mandatory.

A COMPUTERISED DATABASE FOR THE DIAGNOSIS OF RARE DYSMORPHIC SYNDROMES

It has become increasingly clear that the memory afforded by microcomputers must be harnessed to help the clinician remember the almost infinite number of possible combinations of dysmorphic features which constitute recognisable syndromes.[1,5] Before attempting to utilise modern technology, the clinician must identify what he/she expects from the programme. The following priorities were decided upon:
a) that the main function of the computer programme would be to recall, rapidly, all the syndromes previously recorded in the literature;
b) that the experienced dysmorphologist would want, from the computer, a small and manageable list of syndromes featuring a combination of handles;
c) that it would not be necessary for the computer to make the diagnosis, but only to suggest to the clinician the possibilities that he/she cannot readily call to mind;
d) that references are essential so that pictures of the patient under consideration can be compared with those in the original article.

CONSTRUCTION OF THE DATABASE

A list of approximately 1,400 dysmorphic features covering every system in the body was constructed under general headings; for instance, nose, eyes, heart (see Table 1). Each region was divided into 3 levels. The first is a general one (e.g., something wrong with the nose), the second is more specific, and the third level even more detailed (see Table 2). This strategy copes with the variation known to occur within syndromes. Some syndromes might be lost if the search were to be too specific. For instance, a cleft palate might be present in a syndrome with varying frequency, and in some patients there might only be a high palate. By searching on 'something wrong with the palate' rather than 'cleft palate' this problem of variation can be overcome.

TABLE 1

Code selection

Select a code

1) Build
2) Stature
3) Cranium
4) Hair
5) Forehead
6) Ears
7) Eyes (globes)
8) Eyes (general)
9) Nose
10) Face
11) Mouth
12) Oral region
13) Teeth
14) Voice
15) Neck
16) Back and spine
17) Thorax
18) Abdomen
19) Pelvis
20) Genitalia
21) Urinary
22) Upper limbs
23) Hands
24) Nails
25) Lower limbs
26) Feet
27) Blood vessels
28) Endocrine
29) Haematol/immunology
30) Muscles
31) Joints
32) Neuro
33) Skeletal
34) Skin

In general, it is better to combine 1 specific feature (e.g., absent hair) with a general category (e.g., mental retardation), if that is what the patient has, even if that patient also has agenesis of the corpus collosum. Because this last-mentioned feature might be either variable or shared by many other syndromes, a general search is better than a specific one. However, the system is versatile enough to do many searches within a short time, and different strategies could be tried before making a final diagnosis or before deciding that such a syndrome has not been reported previously in the literature. The system also allows the

TABLE 2
The 3-level code for 'nose'

A search can be general (Level 1) or specific (Level II).

Level I	Nose General	
Level II	Alae nasi	general
	Nasal bridge	general
	Nasal columella general	
	Nares	general
	Septum	general
	Tip	general
Level III	Depressed/flat nasal bridge	
	High/prominent nasal bridge	
	Thin nasal bridge	
	Wide nasal bridge	

clinician to ask for 1 specific rare malformation, but unless it is particularly rare the list would probably be too long for the clinician to look up all the references.

The success or failure of the system also depends on the quality of the data and every effort has been made to update these data regularly, and to ensure that the system is totally comprehensive. There is also a facility whereby the clinician can, within seconds, look at all the features of every syndrome in the database (at present there are 1,500 syndromes that can be searched). The database does not, at this stage, include chromosomal syndromes, but this is the subject of a separate project in collaboration with Dr. Schintzel.

In its present form, no attempt has been made to place the possible diagnoses in order of preference. This can only be done if the individual features are weighted in order of importance. Synophris, for example, might be considered an important feature for the diagnosis of the Cornelia de Lange syndrome, and a score out of 10 could be given according to its presence in the over 100 reported cases in the literature. This could lead to an assessment of the likelihood of the syndrome in a child with many of the other features (e.g., microcephaly, short stature, thin upper lip and hand malformation) but not synophris. Unfortunately this is not possible for the vast majority of dysmorphic syndromes, by virtue of

TABLE 3
Syndrome programme

C	Code entry/search
R	Look at references
S	Look at syndrome features
Q	Quit

their rarity. Indeed, some have been reported only once previously in the literature. In addition, even the well known syndromes are variable, so that there are very few clinical features which must always be present before a diagnosis is made. The programme runs on a microcomputer with a storage capacity of about 10 megabytes and a hard disc is usually necessary. This amount of memory exists in many easily available (and relatively cheap) computers. The programme utilises dBase II or dBase III and it is designed especially to extract from the data various combinations of dysmorphic features. An example is given in Tables 3 and 4. If references on, say, the autosomal recessive form of adrenoleuco-dystrophy are needed, this is easily accomplished (see Table 5).

Computers can also be used to store data on undiagnosed cases which can then be compared to define new syndromes. Since this approach requires input from many clinicians, informal dysmorphology meetings, where paediatricians and clinical geneticists discuss difficult or unusual cases, are a useful source of information. Such a group meets regularly at the Institute of Child Health in London.

TABLE 4

Syndrome search 1.1.80

Selection criteria:

Pig. ret (excl. RP)/Chorioretinitis AND Mental retardation

1) ADRENOLEUKODYSTROPHY – POSSIBLE AR TYPE
2) ARTERIO-HEPATIC DYSPLASIA (ALAGILLE)
3) CAO (1977) AGENESIS OF CORPUS CALLOSUM, MICROCEPHALY
4) COCKAYNE
5) COHEN
6) FURUKAWA (1968) MUSCLE ATROPHY-ATAXIA-RET PIGMENT- DIABETES
7) HALAL (1983) MICROCEPHALY-CLEFT PALATE-RETINAL PIGMENTATION
8) HAPPY PUPPET
9) INFANTILE REFSUM
10) LEBER AMAUROSIS
11) LEVIC (1975)
    OPHTHALMOPLEGIA-MR-LINGUA SCROTALIS
12) MAINZER-SALDINO-RETINAL DYSPL. RENAL DEFECTS. SKELETAL ANOMS
13) MICROCEPHALY AND CHORIO-RETINAL DYSPLASIA (AD)
14) MULIBREY NANISM
15) OCULO-RENAL-CEREBELLAR SYNDROME
16) PFEIFFER (1973) COCKAYNE-LIKE SYNDROME
17) PHILLIPS (1979) THORACIC DYSPLASIS, RETINAL APLASIA
18) SABINAS (TRICHO-ONYCHO-OLIGOPHRENIA)
19) SJÖGREN-LARSSON
20) SYBERT (1980) SH STATURE-ABNORMAL SKIN PIGMENTATION-MR
21) TRICHOMEGALY WITH RETINAL DEGENERATION
22) VAN DEN BOSCH -X-LINKED MR PLUS
23) WRINKLY SKIN SYNDROME

TABLE 5
Example of references

Mnemonic: ADRENOLEUKODYS (AR)
Author (s): BENKE PJ, REYES PF, PARKER JC
Title:NEW FORM OF ADRENOLEUKODYSTROPHY

Journal: HUM GENET   Vol/Page:58:204-208 Date:1981
Comments: TWO SIBS (IM + IM)

Mnemonic: ADRENOLEUKODYS (AR)
Author (s): KELLY RI, MOSER HW
Title: HYPERPIPECOLIC ACIDEMIA IN NEONATAL ADRENOLEUKODYSSTROPHY

Journal: AM J MED GENET   Vol/Page:19:791-795 Date;1984
Comments:GOOD REVIEW OF BIOCHEMISTRY

Mnemonic: ADRENOLEUKODYS (AR)
Author (s) MOSER HW, MOSER AE, SINGH I, O'NEILL BP
Title: ADRENOLEUKODYSTROPHY: SURVEY OF 303 CASES. BIOCHEMISRY,
DIAGNOSIS AND THERAPY.

Journal: ANN NEUROL   Vol/Page;16;628-41 Date;1984
Comments:GOOD REVIEW

The correct diagnosis of rare dysmorphic syndromes cannot be dismissed as 'stamp collecting'. In these days of increasingly sophisticated prenatal tests, accurate recurrence risks must be assessed for the correct management of future pregnancies. Every effort should be made to establish a diagnosis in all cases, and this process should include searches of published reports and access to computer systems and panels of experts.

ACKNOWLEDGEMENT

The above project was done with Dr. Robin M. Winter, Consultant in Clinical Genetics, Northwick Park Hospital, Watford Road, Harrow, Middlesex, and was supported by Action Research for the Crippled Child.

BIBLIOGRAPHY

1 Buyse, M.: 1980, *Centre for birth defects information services.* Birth Defects, *16* (5), 83.
2 Gorlin, R. J.: 1980, *Diagnosis of craniofacial anomalies: subjective evaluation/gestalt,* Birth Defects, *16*, (5), 35.
3 Pinsky, L.: 1978, *The syndromology of anorectal malformation (atresia, stenosis, ectopia),* Amer. J. Med. Genet., *1*, 461.
4 Smith, D. W.: 1982, *Recognisable Patterns of Human Malformation,* 3rd ed., W. B. Saunders, Philadelphia.
5 Winter, R. M., Baraitser, M. and Douglas, M. J. A.: 1984, *Computerized data-base for the diagnosis of rare dysmorphic syndromes,* J. Med. Genet., *21*, 121.

GENETIC CAUSES OF MENTAL RETARDATION IN INDIA

Genetic disorders are generally believed to be of little consequence in developing countries like India, in view of the high infant mortality due to infectious and parasitic diseases. However, the operation of a number of factors in developing countries (such as a huge population, a high birth rate, large family size and preference for consanguineous marriages in many communities) leads to the presence of a large number of patients with genetic diseases.[28] This is especially so in the case of mental retardation. Almost 27% of brain growth takes place during intrauterine life[6], when, in view of the protective environment, genetic factors play a major role. This paper examines the genetic causes of mental retardation in India, and emphasises that genetic services need to be provided at least in the university hospitals, in order to fulfill the objectives of "Health for All by the Year 2000 A.D."

PREVALENCE OF MENTAL RETARDATION IN INDIA

The results of surveys specifically carried out to determine the prevalence of mental retardation, by measuring the I.Q. of persons in the general population, are summarised in Table 1. The mean frequency derived from 3 surveys is 26.9 per 1,000. However, this figure should be extrapolated to the whole of India with caution, because the country is vast and the population in different parts is heterogeneous ethnically, socially and culturally.

TABLE 1
Population-based surveys of mental retardation (MR) in India

Investigators	Place	No. of persons screened	Age group	MR per 1000	
				IQ < 70	IQ < 50
Verma[37]	Nagpur	30,326	All ages	30.0	
			8–15 yr	42.0	
Gupta and Sethi[10]	Lucknow	8,583	All ages	23.3	
Narayanan*[17]	Villages near Bangalore	10,700	All ages		3.4
			5–9 yr		12.4
		3,077	All ages		6.8
Subramanya[24]	Bangalore district	1,498	Children 3–14 yr	27.4	

* Screened only for moderate and severe mental retardation

99

E.K. Hicks and J.M. Berg (eds.), The Genetics of Mental Retardation, 99–106

GENETIC CAUSES OF MENTAL RETARDATION

Analysis of 4 hospital-based studies,[1,13,22,30] in which unselected groups of patients were examined to define the aetiology of mental retardation, revealed that 42% of the cases were due to genetic factors (range 29.6 to 53%).

We carried out a comprehensive analysis of 322 patients with mental retardation attending the Genetics Unit at our hospital from April 1981 through March 1984. Formal IQ testing was done (Gesell's screening test in those below 5 years of age, and the Stanford-Binet test in those above 5 years). Cytogenetic studies (including banding), ophthalmologic examination, radiographs of the skull, and chromatographic and chemical tests on the urine were carried out in every case. Qualitative and quantitative tests for mucopolysaccharides in the urine, serum antibodies to rubella, cytomegalovirus and toxoplasmosis, evaluation of thyroid status, electroencephalography and CAT scan studies were done whenever necessary. After excluding 47 cases due to environmental factors, the remaining 275 were classified in genetic terms (Table 2).

TABLE 2
Genetic aspects of mental retardation in Delhi* and Bombay**

Factor	Delhi		Bombay	
	No. of patients	%	No. of patients	%
Chromosomal abnormalities	107	38.9	248	35.0
Metabolic disorders	34	12.4	54	7.6
Identifiable syndromes	21	7.6	146	20.6
Undifferentiated	113	41.1	261	36.8
Total	275	100.0	709	100.0

* The data or Delhi are from our own studies described in greater detail in the text
** The data for Bombay are from Ambani[3]

CHROMOSOMAL DISORDERS IN INDIA

The frequency of Down syndrome at birth in different parts of India is shown in Table 3. These studies relied on clinical observation, although confirmation was sought by cytogenetic analysis whenever possible. The overall frequency obtained (1.13 per 1,000 births, or 1 per 882) is probably an underestimate, as cases of Down syndrome may remain undiagnosed at birth.[11] In the current population of India (750 million), and with a birth rate of 32 per 1,000, every year 27,120 cases of Down syndrome are born. This works out to 3 cases of Down syndrome born every hour!

TABLE 3
Down syndrome at birth in India

Place	Investigators	No. of births	No. with down syndrome	Per 1000	One per
Delhi	Ghosh & Bali[7]	4,353	6	1.38	726
Delhi, AIIMS	Verma & Singh[36]	9,510	11	1.16	865
Patna	Khanna & Prasad[14]	5,376	7	1.30	768
Bombay	Aiyar & Agarwal[2]	10,000	16	1.60	894
	Tiberwala & Pai[26]	12,360	9	0.73	
Madras	Chandra and Harilal[5]	24,192	25	1.03	968
Trivandrum	Sagunabai et al.[21]	7,167	6	0.84	1195
Ajmer	Gupta et al.[4]	2,145	2	0.93	1073
Total		75,103	82	1.09	916

Studies on X-chromatin in the amniotic membranes of newborns[32] revealed 3 X-chromatin positive males among 1,546 boys (frequency 1.94 per 1,000 males). This is significant because X-chromatin positive males have an almost 10x greater risk of mental retardation as compared with normal males.

TABLE 4
Chromosomal studies in Down syndrome in Delhi (n = 350)

Karyotype	Maternal age at conception			
	Less than 3 yr		30 yr or more	
	No.	%	No.	%
Trisomy 21	253	94.4	81	98.8
Translocation				
D/G	7	2.6	–	–
G/G	5	1.9	1	1.2
Mosaic	2	0.7	–	–
Others	1	0.4	–	–
Total	268	100%	82	100%

CHROMOSOMAL STUDIES IN DOWN SYNDROME

An analysis of 350 cases of Down syndrome at our genetic laboratory is presented in Table 4. It is seen that 4.5% of cases had a translocation karyotype when the maternal age at conception was less than 30 years.

MATERNAL AGE IN DOWN SYNDROME

The proportion of cases of Down syndrome born to mothers aged 35 years or more at conception has fallen considerably in India in the past 15 years.[33] This diminution has paralleled the decrease in the control population (women bearing a normal infant) (Table 5). The data in the 1980–85 period are compared with

TABLE 5
Maternal age at conception in cases of Down syndrome in Delhi (% of all women)

Maternal age group (yr)	1960–1979		1980–1985	
	Control	Study	Control	Study
n =	59,479	615	1,226	514
15–19	11.8	5.6	4.6	9.7
20–24	35.5	20.6	37.2	32.5
25–29	28.8	22.8	41.1	30.2
30–34	16.7	18.0	9.5	13.8
35–39	6.2	23.6	2.8	11.1
40–47	1.2	9.4	0.8	2.7

those in the period 1960–1979. It is clear that the proportion of cases of Down syndrome born to women 35 years or more has decreased from 33% to 13.8%, while in the control population it has decreased from 7.4% to 3.6%. This shows that the family planning programmes have had an impact, as less women are conceiving at older ages. This trend is similar to the one observed in many developed countries, although the reasons are probably different.

DOWN SYNDROME AND HIGH BACKGROUND RADIATION IN KERALA

In the Quilon district of Kerala, the Chavara-Nandakara area has a high background radiation (10–20 times higher than the usual background radiation elsewhere in India). This is due to monazite in the soil which contains thorium. A population of 12,000 persons in this area, and about 6,000 in a control area (Purrakade district) higher up on the coast, was examined. There were 12 cases of Down syndrome in the study area, as compared with none in the control area. The cases of Down syndrome, their parents and the workers employed in the factory dealing with monazite also showed higher incidence of chromosome-type aberrations, but no increase in chromatid-type aberrations.[34] Eleven of the 12 cases of Down syndrome occurred in women over the age of 30 years at the time of conception. It was concluded that the high incidence of Down syndrome was related to the effect of high background radiation exacerbating the effect of advanced maternal age at conception.[15,35]

NEURAL TUBE DEFECTS IN INDIA

An analysis of 515,500 births from various parts of India revealed a frequency of anencephaly and spina bifida of 2.5 and 1.5 per 1,000 births (comprising live and stillbirths) respectively, and 4 per 1,000 for both together.[27,29] However, it is remarkable that the incidence was very high in North India (Punjab 1:116 births, Rajasthan 1:145 births, and Delhi 1:212 births). The high incidence among the Sikhs had already been recognised in studies among Sikh immigrants in Singapore[23] and the United Kingdom.[16] However, the present analysis showed that the incidence was also high among other communities in North India.[27,29] In comparison, the frequency in West India (Bombay) was 1:450 births, in South India (Tamil Nadu) 1:330 and in East India (Calcutta) 1:1100 births.

We have set up a comprehensive screening programme for neural tube defects. Alpha feto-protein (AFP) is measured in maternal serum by an enzyme-linked immuno-assay (ELISA), amniotic fluid AFP is determined by immunoelectro-phoresis and acetyl-cholinesterase is detected by gel electrophoresis.[8,12] A vitamin supplementation programme during the periconceptional period is also being tried out in the North Indian population.

METABOLIC DISORDERS IN INDIA

Our comprehensive analysis of 275 cases of mental retardation due to genetic causes in Delhi (Tables 2 and 6) showed metabolic disorders as the basis of the retardation in 34 (12.4%). Mucopolysaccharidoses were the largest group (20), followed by 3 cases each of hypothyroidism and galactosaemia, 2 each of phenyl-ketonuria, degenerative brain disease, and Tay-Sachs disease, and 1 each of Lowe syndrome and glycogen storage disease.

TABLE 6
Biochemical abnormalities in mental retardation in India

Plate	Investigators	No. due to genetic causes	Biochemical abnormalities	
			No.	%
Delhi	Verma[30]	275	34	12.4
Delhi	Sinclair[22]	139	9	6.5
Bombay	Patel[18]	460	18	3.9
Bombay	Ambani[3]	295	16	5.4
Bangalore	Thomas[25]	328	9	2.7
Lucknow	Agarwal[1]	273	6	2.2
Varanasi	Subramanyan[24]	70	5	7.1
Vellore	Joshua[13]	303	59	19.5

A review of studies in India shows that biochemical disorders as a cause of genetically determined mental retardation range from 2.2% to 19.5% (Table 6), depending upon the availability of facilities for biochemical analysis, as well as the interests of the investigators. While mucopolysaccharidoses were the commonest in our study in Delhi, in Vellore[13] a high prevalence of sphingolipid disorders was found, and in Karnataka[20] a high prevalence of phenylketonuria was observed.

The only study of mass screening of newborns in India has been carried out in Karnataka[4]; 59,840 blood samples from newborns have been analysed for aminoacid disorders. Four cases of hyperphenylalaninaemia were detected; 2 of these had phenylketonuria, 1 had a transient abnormality and 1 was lost to follow up. There were 17 infants showing tyrosinaemia, of whom 1 died, 9 were normal subsequently and the rest were lost to follow up. There were 3 cases of branched chain aminoaciduria of whom 1 died. Three infants were identified to have histidinaemia.

CONSANGUINITY AND MENTAL RETARDATION

Two patterns of consanguinity are seen in India. In the south, where Dravidian languages are spoken, almost 20–50% of marriages, irrespective of religion, are consanguineous. Most of these are between first cousins, but about 5–10% are between uncle and niece. In North India, where Aryan language is spoken, consanguineous marriages take place almost exclusively among Muslims, nearly always between first cousins.

A number of hospital-based studies on mentally retarded children have shown a higher incidence of mental retardation among the offspring of consanguineous marriages. We carried out a community study of 1,494 couples and their 3,206 children in Pondicherry, South India.[19] Of all marriages 55.4% were consanguineous: 31.4% in which the spouses were first cousins, 20.2% uncle-niece, and 3.8% related more distantly than first cousins. The prevalence of mental retardation was 2.4% among 2,012 children of consanguineous parents, as against 1.8% among 1,194 children of non-consanguineous parents. There was a higher frequency of autosomal recessive disorders (primary microcephaly, Hurler syndrome, and Bardet-Biedl syndrome) in the consanguineous unions, while there was no increase in autosomal dominant, X-linked or chromosomal disorders.

PREVENTION AND MANAGEMENT

It is apparent that for prevention and management of mental retardation in India provision of genetic services should constitute a high priority. At present it is advisable to locate the genetic centres in university hospitals. These centres should have facilities for cytogenetic and biochemical tests, and for genetic counselling and prenatal diagnosis.

An extensive health care infrastructure has been established throughout the country, operating through primary health centres and district hospitals. These can ensure recognition of patients with genetic diseases in the peripheral villages, and their referral to district hospitals where the paediatrician/internist can counsel the simple cases. The more complicated cases can be referred to the genetic centres at the university hospitals.

Initially the focus should be on the investigation and management of affected families. At present screening should be restricted to high-risk populations, although pilot studies on screening of newborns for genetic disorders may be started on a research basis, in selected centres in different parts of the country, to gather data on incidence of various disorders. Widespread screening of newborns must await the control of common infectious diseases, such as neonatal tetanus and diarrheal disorders, so that the infant mortality rate is brought down below 60 per 1,000, as envisaged in the 'Health for All by the Year 2000 A.D. plans of the government of India.

BIBLIOGRAPHY

1 Agarwal, S. S.: 1984, *Genetic causes of mental retardation in Lucknow – a multi-centric study*, Report submitted to Indian Council of Medical Research, New Delhi.
2 Aiyar, R.R. and Agarwal, J.R.: 1969, *Observations on the newborns – study of 10,000 consecutive live births*, Indian Pediat., *6*, 729.
3 Ambani, L. M.: 1984, *Clinical, biochemical, and cytogenetic studies in mental retardation*, Indian J.Med. Res., *79*, 384.
4 Bittles, A. H., Devi, A. R., Rao, S. V. and Rao, N. A.: 1982, *A newborn screening programme for the detection of aminoacid disorders in South India*, Biochem. Rev., *52*, 21.
5 Chandra, P. and Harilal, K.T.: 1978, *Congenital malformations in Madras: a study of 24,192 consecutive births*, In: I. C. Verma, ed., Medical Genetics in India, Aruoma Enterprises, Pondicherry, 47–54.
6 Dobbing, J.: 1974, *The later development of the brain and its variability*, In: J. A. Davis and J. Dobbing, eds., Scientific Foundations of Paediatrics, William Heinemann Medical Books Ltd., London, 565–576.
7 Ghosh, S. and Bali, L.: 1963, *Congenital malformations in newborn*, Indian J. Child Hlth., *12*, 448.
8 Gupta, A., Verma, I. C., Aurora, N. K., Kumar, R., Bhargava, V. and Ghai, O. P.: 1982, *Maternal serum alpha fetoprotein assay using ELISA in normal pregnancy*, Indian J. Med. Res., *76*, 843.
9 Gupta, B. M., Mathur, H. C. and Sharda, D.C.: 1971, *A study of congenital malformations in central Rajasthan (Ajmer)*, Arch. Child Hlth., *13*, 30.
10 Gupta, S. C. and Sethi, B. B.: 1970, *Prevalence of mental retardation in Utter Pradesh*, Indian J. Psychiat., *12*, 264.
11 Hall, B.: 1964, *Mongolism in newborns – a clinical and cytogenetic study*, Acta Pediat., suppl 154.
12 Jacob, T. T., Verma, I. C. and Hingorani, V.: 1977, *Study of alpha-fetoprotein in amniotic fluid and its application in prenatal diagnosis of neural tube defects*, Indian J. Med. Res., *66*, 236.
13 Joshua, G.E.: 1974, *Mental retardation in children – 1. An etiologic study of 303 cases*, Indian Pediat., *99*, 47.
14 Khanna, K. K. and Prasad, L. S.: 1967, *Congenital malformations in newborn*, Indian J. Pediat., *34*, 63.
15 Kochupillai, N., Verma, I.C., Grewal, M.S. and Ramalingaswami, V.: 1976, *Down syndrome and related abnormalities in an area of high background radiation in coastal Kerala*, Nature, *262*, 60.
16 Leck, I.: 1969, *Ethnic variations in the incidence of malformations following migration*, Brit. J. Prev. Soc. Med., *23*, 166.

106

17 Narayanan, H. S.: 1981, *A study of the prevalence of mental retardation in Southern India*, Internat. J. Ment. Hlth., *10*, 28.

18 Patel, Z.: 1984, *Genetic causes of mental retardation in Bombay – a multi-centric study*, Report submitted to Indian Council of Medical Research, New Delhi.

19 Puri, R. K., Verma, I. C. and Bhargava, I.: 1978 *Effects of consanguinity in a community in Pondicherry*, In, I. C. Verma, ed., Medical Genetics in India, vol 2. Auroma Enterprises, Pondicherry 129–140.

20 Rao, B. S., Subash, M. N. and Narayanan, H. S.: 1978, *Biochemical screening of cases of mental retardation in Bangalore*, In, I. C. Verma, ed., Medical Genetics in India, vol. 1, Auroma Enterprises, Pondicherry, 93–96.

21 Sagunabai, N. S., Mascarene, M., Syamalan, K. and Nair, P. M.: 1982, *An etiologic study of congenital malformations in the newborn*, Indian Pediat., *19*, 1003.

22 Sinclair, S.: 1972, *Aetiologic factors in mental retardation – a study of 470 cases*, Indian Pediat., *9*, 391.

23 Stevenson, A. C., Johnston, H. A., Stewart, M. P. and Golding, D. R.: 1966, *Congenital malformations – a report of a study of series of consecutive births in 24 centres*, Bull. W.H.O., Suppl. to Vol. 34.

24 Subramanya, B.: 1983, *An epidemiological study of mental retardation in rural children*, M.D. thesis submitted to the University of Bangalore, Bangalore.

25 Thomas, I. M.: 1984, *Genetic causes of mental retardation in Bangalore – a multicentric study*, Report submitted to Indian Council of Medical Research, New Delhi.

26 Tiberwala, N. S. and Pai, P.M.: 1974, *Congenital malformations in the newborn period*, Indian Pediat., *6*, 729.

27 Verma, I. C.: 1978, *High frequency of neural tube defects in India*, Lancet, *1*, 879.

28 Verma, I. C.: 1978, *Genetic disorders in India*, In, I. C. Verma, ed., Medical Genetics in India, vol. 1, Auroma Enterprises, Pondicherry, 5–18.

29 Verma, I. C.: 1978, *Neural tube defects in India*, In, I. C. Verma, ed., Medical Genetics in India, vol. 1, Auroma Enterprises, Pondicherry, 33–34.

30 Verma, I. C.: 1984, *Genetic causes of mental retardation in Delhi – a multi-centric study*, Report submitted to Indian Council of Medical Research, New Delhi.

31 Verma, I. C.: 1986, *Genetic disorders need more attention in developing countries*, World Health Forum. W.H.O., Geneva, *7*, 69.

32 Verma, I. C., Bawa, B., Ghai, O. P., Hingorani, V. and Guha, D. K.: 1974, *Epidemiology of X-chromatin aberrations in newborns in Delhi*, Indian J. Med. Res., *62*, 676.

33 Verma, I. C., Chordiya, A. and Sundaram, K.: 1987, *Changing pattern of maternal age at conception in Down syndrome in North India*, (in press, Indian J. Med. Res).

34 Verma, I. C., Kochupillai, N., Grewal, M. S., Mallick, G. R. and Ramalingaswami, V.: 1977, *Genetic effects of high background radiation in coastal Kerala (India). 1-Clinical and cytogenetic studies*, In, International Symposium on Areas of High Natural Radioactivity, Pocas de Carlos, Brazil, Academia Brasilera de Genceas, Rio de Janeiro, 186–187.

35 Verma, I. C., Kochupillai, N., Grewal, M. S., Ramachandran, K. and Ramalingaswami, V.: 1977, *Down syndrome in Kerala*, Nature, 267, 728.

36 Verma, I. C. and Singh, M. B.: 1975, *Down syndrome in India*, Lancet, *1*, 1200.

37 Verma, S. K.: 1968, *Research and rehabilitation project of mental retardation at Nandanvan, Nagpur*, Indian J. Ment. Retard., *1*, 49.

M. F. NIERMEIJER

POSTNATAL DIAGNOSIS – THE NEXT DECADE

New diagnostic technologies, improvements in classification, and preventive strategies have widened the options available in genetic counselling. This has not only increased the responsibility of making a correct diagnosis, but has also added to its complexity. By 1985 there were ± 300 restriction fragment length polymorphisms (RFLPs) known to be distributed among all human chromosomes, and this list of potentially informative markers for genetic diseases will expand at an increasing rate.[7,21] Will the clinical geneticist of the future only be required to 'measure the RFLP-status' of his consultands and counsel them on DNA studies in future pregnancies?

A geneticist is confronted not only with the development of DNA technology and its implications, but also with sharper delineation of syndromes.[22] One is acutely aware that a patient is diseased as a result of a malformed protein rather than from the single presence of a certain gene mutation. The contribution of DNA technology and several hundred potential markers should be seen in light of the currently known $\pm 4{,}000$ Mendelian disorders, and the estimated $\pm 50{,}000$ structural human genes, together with their potential associated abnormalities and related syndromes.[12] These simple comparisons make it very clear how important clinical diagnosis and clinical genetic analysis is and will remain for the next decades.

This presentation describes some of the discrepancies which arise between advanced technology, together with the many types of mental retardation and congenital handicaps about which little is understood. The focus will successively be on chromosomal disorders, monogenetic (autosomal dominant, autosomal recessive and X-linked) disorders, multifactorial diseases and interactions of genes and the environment.

CHROMOSOMAL DISORDERS

In chromosomal disorders (1:200 newborns), improvements in cytogenetic diagnostic techniques (microcytogenetics, allowing high resolution analysis of smaller chromosomal defects[24]) have facilitated greater insight into some cases of aniridia, retinoblastoma, and the Prader-Willi syndrome.[2] These techniques are not, however, generally available as yet, nor are they easy to apply. Advances in delineation of chromosomal instability syndromes (e.g., Fanconi syndrome, Bloom syndrome) have been made, enabling prenatal diagnosis in many of the clearly identified families.

The delineation of the X-linked mental retardation with the fragile site on the X-chromosome (one of the most frequent individual forms of mental retardation)

E.K. Hicks and J.M. Berg (eds.), The Genetics of Mental Retardation, 107–114
© *1988 by Kluwer Academic Publishers.*

has greatly helped in making a diagnosis, but has not solved the problem of carrier detection (an important fraction of the carriers not being detectable). The variable expression of this marker also made prenatal monitoring of at-risk pregnancies less precise than desirable. Future work on DNA-markers may, however, improve the situation[15] and Fryns (this volume). This is a clear example of how technological improvements aided the diagnostic classification of a large group of mentally retarded males (and some carrier females), but have not as yet solved the ensuing problem of carrier detection and prenatal diagnosis in at-risk pregnancies.

AUTOSOMAL DOMINANT DISORDERS

In autosomal dominant Mendelian disorders there is a fascinating development along separate lines. Studies on the variability of the many autosomal dominant inherited disorders (from neurofibromatosis to von Hippel Lindau's disease, and from polycystic kidneys to polyposis coli[12]) have attracted new interest. Firstly, to establish reasonable standards for clinical diagnosis, necessary to the development of an acceptable degree of diagnostic accuracy for genetic counselling in highly variable disorders. The use of ultrasonography as a diagnostic tool for presymptomatic diagnosis of adult polycystic kidney disease (autosomal dominant) is an example. Moreover, when DNA-marker studies are to be used for presymptomatic diagnosis, and even prenatal diagnosis, precise data on the natural course of a disease are necessary in order to obtain accurate prior risk estimates about a person. Particularly in disorders with late onset, data on the natural variability contribute to a proper interpretation of data when using DNA-linked marker analysis in the younger generations (e.g. in families at risk for adult polycystic kidney disease, using the RFLP on chromosome 16).[18]

An increasing number of dominant diseases are being 'mapped' to certain chromosomal regions. This will lead to the use of RFLP-analysis as a diagnostic marker for a disease where sufficient supportive data can be obtained from the family to permit an informed analysis. Such mapping studies will not only increase the growing need for family studies (and the involvement of relatives in the reproductive options of a few of their family members), but also presents a problem for counselling and decision making when recombination at meiosis may result in a false prediction in, for example, 5% of cases. Since it is predictable that studies of linked RFLP-markers will often be more feasible than diagnosis at the level of the gene mutation itself, we must develop a way of dealing with the uncertainty raised by high-tech tests. It is rather impossible to say that 2% inaccuracy is acceptable, whereas 5% or more is not.[12,13] The question for families or parents is: if one does not accept a 50% risk of recurrence, is one going to accept a 2% risk, or should the option be that 'it should never happen again?' The implications of these choices must be made clear to the parents involved, when linkage studies are being used for prenatal diagnosis.

In studies involving presymptomatic diagnosis on young adult patients, there may not only be the problem of possible inaccuracy, but also the question of how to deal with the burden of knowing one's own fate. If a number of relatives in such families wish to know the risk for their own offspring, the consultands often request that this knowledge not include information about the consultands themselves. In Huntington's disease it has been proposed, and is currently applied in a few centres, to test only for the presence or absence of 1 of the affected grandparental haplotypes (if known) and to offer termination of pregnancy if 1 of these is found. This circumvents the necessity to identify the heterozygous state of the disease in the respective parent of the fetus.[14]

Even if some people may benefit from the previous application, other studies will involve predictive genetic testing to inform people of their own risks. If the particular disorder is amenable to therapeutic intervention, the benefits may outweigh the potential risk. Current discussion of the application of the chromosome 4 gene probe, to detect a polymorphic region closely linked to the gene for Huntington's chorea, reflects most of these problems. Counselling, and above all supportive strategies are needed here. Genetics, through its far reaching implications and new possibilities, is considered to be increasingly 'at the bedside'; nevertheless, the question remains as to who is really caring for the patient, and how should such care be given?[3]

AUTOSOMAL RECESSIVE DISORDERS

In autosomal recessive diseases the number of disorders identifiable by a specific enzymatic or protein defect only slowly increases. The hemoglobinopathies are the exception here, where studies at the protein level have been followed up by detailed knowledge and diagnostic accuracy at the level of the mutations.[8] The aminoacid disorders (from classical phenylketonuria (PKU) to the many variants of other enzyme defects in this area) represent the 'classic example,' in the 1960s, of metabolic diseases identifiable by an enzyme defect. This was followed by the multitude of lysosomal storage diseases. Then the mitochondria and their disorders were identified,[4] and more recently, the smallest of cellular vesicles, the peroxisomes, began to disclose the nature of their genetic defects (e.g., Zellweger syndrome, Refsum disease, adrenoleukodystrophies).[20] New and debated options became availiable, e.g. prenatal diagnosis of PKU by DNA-linkage studies.[23] The perception of the burden of this disease for future children may be the parent's motivation for such a choice. A similar problem exists in adrenal hyperplasia.[16] Testing for disorders with adult-age onset of manifestations (e.g. emphysema, liver disease) became possible with prenatal DNA analysis for alpha$_1$-antitrypsin deficiency.[9] The story of the development of these discoveries shows the complexity of this field.

The rapid progress in single areas sharply contrasts with the slower progress in better understanding of many malformation complexes, manifesting as autosomal recessive diseases, as well as the slow progress in understanding relatively common recessive diseases such as cystic fibrosis (CF). CF represents an important example of an asychronous development by the detection of a number of polymorphic regions (RFLP'a) closely linked to the CF gene on chromosome 7; whereas neither the genetic defect itself, nor any specific traditional diagnostic tests are known or available. Even if the chromosomal localisation of the CF gene may eventually lead to the clarification of the basic defect, there will be a period in which we use 'imperfect understanding but a useful marker' to do prenatal diagnosis for CF in early pregnancy.[5] Is this half-technology, or are we offering people only half a hand in their confrontation with the full burden of human suffering?

Dysmorphology has silently, but effectively, kept its place in the field of recessive diseases. Since the majority of recessive diseases are not identifiable by specific laboratory techniques, pattern recognition and diagnostic curiosity are indispensable tools in these often rare disorders with their high risk of recurrence. The development of data bases for syndrome identification (see Baraitser, this volume) will hopefully be of some help in unravelling the anticipated complexity and diversity, assuming that there are $\pm 50,000$ gene loci, with a multitude of alleles at each locus.

The options for parents with a recurrence risk for a syndromal autosomal recessive disorder, which involves both mental retardation and multiple structural abnormalities, have greatly increased through refined prenatal ultrasound imaging. This permits early detection (or exclusion) of many of the more serious forms of an ever growing list of syndromes. The geneticist has a task to study the variability of recessive syndromes, particularly the intrafamilial variability. Such information is essential in counselling, and is only available for a few syndromes such as Meckel's syndrome.[6]

X-LINKED DISORDERS

The number of disorders localised to the X chromosome has steadily grown. Many of these diseases are associated with retardation and have traditionally been identified by their peculiar pedigree pattern. A cytogenetic marker (fragile site on the distal long arm of the X chromosome) was established for the clinically rather aspecific X-linked mental retardation syndrome. This type of retardation is believed to comprise 20% of the mental retardation in males. Its tremendous importance for diagnosis nevertheless leaves a number of uncertainties for carrier detection and prenatal diagnosis (discussed above).

The multitude of genes and markers mapped to several regions of the X chromosome (even if not informative for the X-linked fragile site syndrome), has opened fascinating new options for carrier detection and prenatal monitoring. Many women who are carriers of, for example, Duchenne's muscular dystrophy, hemophilia, adrenoleukodystrophy, and ornithine transcarbamylase deficiency, now have healthy sons after chorionic villi biopsy and RFLP analysis.[1] This is a dramatic improvement over prenatal sex determination and abortion of all male fetuses (the only option available for carriers until a few years ago).

Moreover, not only has carrier detection become reliable, but other women could be told that they were not carriers (which had previously been impossible, as most of the earlier techniques were influenced by X-inactivation). Many of these options are dependent upon the cooperation of affected and unaffected relatives to contribute blood and skin specimens for DNA analysis. Apart from the logistic problems for the relevant laboratory (especially when the consultand is already pregnant), and the desirability of storing cell cultures from affected probands for later comparison and reference, there may be emotional and ethical problems in obtaining participation of all concerned relatives. Data handling and storage, as well as control of confidentiality, should be organised such that future generations of these families can profit from modern diagnostic capability. In some families with X-linked diseases, the problem may arise that the potential risk for the disease is unknown to young female relatives, because the affected relative died years, or generations, before they were born. Information to families and physicians should be disseminated such that its loss can be prevented.

MULTIFACTORIAL DISEASES OR MALFORMATION COMPLEXES

In multifactorial diseases or malformation complexes, the interaction of multiple genetic and environmental factors is still largely not understood. If 2–3% of newborns have a malformation (e.g. spina bifida, cardiac defect, facial clefting, limb position defect, intestinal malformation) associated with this causative mechanism, it is usually assumed that the recurrence risk is small (low percentage) in most 'nonsyndromic' cases (i.e. when the malformation only involves a single organ). Methodologies for prenatal monitoring, involving alpha-fetoprotein analysis of the amniotic fluid and advanced ultrasound imaging, have become available for parents with a recurrence risk of, for example, a neural tube defect, a structural heart defect, and limb reduction defects.

ENVIRONMENTAL INTERACTIONS

The understanding of the relationship between genetic and environmental interactions has become intensified as a result of an increasing concern about teratogenic factors. The maternal use of the anti-epileptic valproate was found to be associated with a 1–2% risk for spina bifida in her offspring, and this could be confirmed in animal studies.[11] In this case, the value of birth defect monitoring registries was also substantiated. Although vitamin A was traditionally known as a teratogen in animal studies, it took until 1985 before the spectrum of defects in the craniofacial, cardiac and central nervous system were fully described, following the use of retinoic acid, the potent vitamin A analogue.[10] A unique feature is the very long storage in fatty tissue of retinoic acid and its metabolites, resulting in a teratogenic potential lasting for months after discontinuation of therapy for acne, psoriasis, or other desquamating skin disorders.

The fetal alcohol syndrome is of great concern because of its potential for

112

causing non-specific mental and growth retardation, which is often not diagnosed because of the absence of specific facial features or malformations.[17] Here, the options for improving prevention are only slowly being implemented.

Clinical genetic centres are increasingly being used for information and consultation about potential teratogenic and mutagenic assaults during and prior to pregnancy. The identification of such assaults requires familiarity with a multitude of chemical and pharmaceutical substances, in addition to access to relevant data.[19]

CLINICAL GENETICS, DIAGNOSIS AND PREVENTION

Even if a 'certain' diagnosis is not always obtainable, it is of paramount importance for parents and probands that an exhaustive evaluation of the problem in their family has been made. The diagnostic facilities used in clinical genetics (dysmorphic evaluation, cytogenetic analysis, urinary metabolite screening, enzyme studies on leukocytes and fibroblasts) permit precise diagnostic classification and appropriate counselling, in most cases. These data may be used by relatives in future decision making. Assuming that especially those at high risk will more often be deterred from having more children, or may use prenatal diagnosis, one may estimate the number of births of affected future children which have been prevented through such precise diagnosis. Results from the 7 clinical genetic centres in the Netherlands for 1984 (Table 1) indicate that, in a country with about 170,000 births per annum, for each of the postnatal cytogenetic, metabolite and enzyme diagnostic categories, the birth of 50–100 subsequent affected individuals was prevented.

In terms of prevention, prenatal diagnosis detects an affected fetus in some 4%

CLINICAL GENETIC CENTERS - THE NETHERLANDS
ANNUAL CASE LOAD (1984)

TYPE STUDY	N Tested	Abnormal N	%	ESTIMATED PREVENTION (+/- PRENATAL DIAGNOSIS)
CYTOGENETICS (postnatal)	5000	1250	25	50–100
URINARY METABOLITES	5000	200	4	50–100
ENZYME TESTS	1500	200	13	50–100
GENETIC COUNSELING	3000			315–630
PRENATAL DIAGNOSIS	3000	120	4	120

of pregnancies studied (at risk for a chromosomal or metabolic disorder, or a neural tube defect). Genetic counselling, especially of complex disorders, is even more important since, out of 3,000 cases, an estimated 300–600 couples will not reproduce in order not to have the affected offspring they otherwise might have had.

At the level of the individual family and their relatives, the effectiveness of genetic counselling is demonstrated by providing either reassurance or timely warning and information, enabling free and informed reproductive decisions. It is to be hoped that with the expansion of knowledge and technology, clinical genetic services will be disposed and equipped to answer the questions of those in need.

BIBLIOGRAPHY

1 Bakker, E., Goor, N., Wrogemann, K., Kunkel, L. M., Fenton, W. A., Majoor-Krakauer, D., Jahoda, M. G. J., Van Ommen, G. J. B., Hofker, M. H., Mandel, J. L., Davies, K. E., Willard, H. F., Sandkuyl, L., S., v. Essen, A. J., Sachs, E. S. and Pearson, P. L.: 1985, *Prenatal diagnosis and carrier detection of Duchenne muscular dystrophy with closely linked RFLPs*, Lancet *1*, 655.

2 Butler, M. G., Meaney, F. J. and Palmer, C. G.: 1986, *Clinical and cytogenetic survey of 39 individuals with Prader-Labhart-Willi syndrome*, Amer. J. Med. Genet., *23*, 793.

3 Craufurd, D. I. O. and Harris, R.: 1986, *Ethics of predictive testing for Huntington's chorea: the need for more information*, Brit. Med. J., *293*, 249.

4 DiMauro, S., Bonilla, E., Zeviana, M., Nakagawa, M. and De Vivo, D.C.: 1985, *Mitochondrial myopathies*, Ann. Neurol., *17*, 521.

5 Farrall, M., Law, H.-Y., Rodeck, C. H., Warren, R., Stanier, P., Super, M., Lissens, W., Scambler, P., Watson, E., Wainwright, B. and Williamson, R.: 1986, *First-trimester prenatal diagnosis of cystic fibrosis with linked DNA probes*, Lancet, *1*, 1402.

6 Fraser, F. C. and Lytwyn, A.: 1981, *Spectrum of anomalies in the Meckel syndrome or: "Maybe there is a malformation syndrome with at least one constant anomaly,"* Amer. J. Med. Genet., *9*, 67.

7 Gusella, J. F.: 1986, *Recombinant DNA techniques in the diagnosis of inherited disorders*, J. Clin. Invest., *77*, 1723.

8 Kan, Y. W.: 1986, *The William Allan Memorial Award Address: Thalassemia: molecular mechanisms and detection*, Amer. J. Hum. Genet., *38*, 4.

9 Kidd, V. J., Golbus, M. S., Wallace, R. B., Itakura, K. and Woo, S. L. L.:1984, *Prenatal diagnosis of α_1-antitrypsin deficiency by direct analysis of the mutation site in the gene*, New Engl. J. Med., *310*, 639.

10 Lammer, E. J., Chen, D. T., Hoar, R. M., Agnish, N. D., Benke, P. J., Braun, J. T., Curry, C. J., Fernhoff, P. M., Grix, A. W., Lott, I. T., Richard, J. M. and Sun, S. C.:1985, *Retinoic acid embryopathy*, New Engl. J. Med., *313*, 837.

11 Lindhout, D. and Meinardi, H.: 1984, *Spina bifida and in-utero exposure to valproate*, Lancet, *2*, 396.

12 McKusick, V. A.: 1986, *Mendelian Inheritance in Man*, 7th ed., Johns Hopkins Press, Baltimore.

13 Merz, B.: 1985, *Markers for disease genes open new era in diagnostic screening*, J. Amer. Med. Ass., *254*, 3153.

14 Merz, B.: 1986, *Screening for cystic fibrosis and Huntington's disease begins*, J. Amer. Med. Ass., *255*, 1826.

15 Opitz, J. M., Reynolds, J. F. and Spano, L. M., eds.: 1986, *X-linked Mental Retardation*, *2*. Amer. J. Med. Genet., *23*, No. 1/2, 737.

16 Philips, I. R. and Shepard E. A.: 1985, *Complementary genes for an adrenal enzyme deficiency*, Nature, *314*, 130.

17 Porter, R., O'Connor, M. and Whelan, J., eds.,: 1984, *Mechanisms of Alcohol Damage in Utero,* Ciba Foundation Symposium 105, Pitman, London.

18 Reeders, S. T., Breuning, M. H., Davies, K. E., Nicholls, R. D., Jarman, A. P., Higgs, D. R., Pearson, P. L. and Weatherall, D. J.: 1985, *A highly polymorphic DNA marker linked to adult polycystic kidney disease on chromosome 16,* Nature, *317,* 542.

19 Schardein, J. L.: 1985, *Chemically Induced Birth Defects,* Marcel Dekker, New York.

20 Schutgens, R. B. H., Heymans, H. S. A., Wanders, R. J. A., Bosch, H. van der, and Tager, J. M.: 1986, *Peroxisomal disorders. A newly recognised group of genetic diseases,* Eur. J. Pediat., *144,* 430.

21 Shapiro, L. S. (moderator): 1986, *New frontiers in genetic medicine,* Ann. Intern. Med., *104,* 527.

22 Weatherall, D. J.: 1985, *The New Genetics and Clinical Practice,* 2nd ed. Oxford University Press, Oxford.

23 Woo, S. L. C.: 1984, *Prenatal diagnosis and carrier detection of classic phenylketonuria by gene analysis,* Pediatrics, *74,* 412.

24 Yunis, J. J. and Lewandowski, R. C.: 1983, *High-resolution cytogenetics,* Birth Defects, Original Article Series *19* (5), 11.

MENTAL RETARDATION SYNDROMES RECOGNISED AFTER BIRTH: PROBLEMS OF DIAGNOSIS AND MANAGEMENT

A wide variety of conditions associated with mental retardation can be recognised or suspected at birth, or later in infancy or childhood. Some are identified by physical examination of the newborn baby or by routine screening tests performed soon after birth. Others are not identifiable at birth by presently available techniques, although the family history may indicate that non-routine tests should be performed on selected infants.

Before the doctor talks with the parents, careful thought must be given to a number of questions:

1. Is this condition usually recognised at birth?

Children with mental retardation syndromes usually present to the doctor in two quite separate ways. First, the syndrome may be recognised at or soon after birth by physical, biochemical or chromosomal features, and the doctor must predict the likely outcome for development and intelligence. It is for this category of child that the following questions are particularly important. Second, the child may present with delayed development/mental retardation, and subsequent investigations may identify a specific cause of genetic, prenatal, perinatal or postnatal origin.

Parents will often want to know – could this have been diagnosed earlier? For a condition recognised at birth, could it have been diagnosed prenatally? For one recognised later, could it have been recognised at birth? Such questions need to be answered honestly but carefully, because the answers could lead to accusations of negligence against other doctors.

2. Is this a condition for which there is definitive diagnosis?

For some conditions associated with mental retardation, a biochemical, chromosomal or histological abnormality is characteristic and diagnostic of that disorder. Provided there have been no technical mistakes, the diagnosis is certain. There may still be room for discussion about the significance of such a laboratory diagnosis. For example, histidinaemia was at one time thought to be always associated with mental retardation, and XYY chromosomes with a psychopathic personality. It is now known that neither of these generalisations is true.

For other conditions, diagnosis rests entirely on physical examination of the child, and therefore there may be a difference of clinical opinion. This is particularly likely to arise with conditions in which the physical features vary greatly from one individual to another. When there is doubt, the opinions of experienced colleagues are valuable.

3. Is this condition invariably associated with mental retardation? How variable is the degree of retardation?

E.K. Hicks and J.M. Berg (eds.), The Genetics of Mental Retardation, 115–120

In some conditions, such as severe congenital microcephaly, severe mental retardation can be predicted with confidence. In others, such as Down syndrome, one can predict significant retardation of variable degree. In yet others the retardation may vary from severe to mild. Finally, there are conditions in which associated retardation is quite common but in which intelligence may be completely normal. The most difficult thing for people to live with is uncertainty. Conditions recognisable in early life which are very variably associated with mental retardation therefore present particularly difficult problems for parent counselling.

4. Is there any effective treatment for this condition?

If effective treatment is available, early diagnosis is urgent. The degree of success of treatments varies according to the condition and the complexity of therapy. Treatment of congenital hypothyroidism with thyroxine is simple, and the results are excellent if started early enough. This is why an effective neonatal screening programme is essential. Dietary treatment of phenylketonuria and surgical treatment of hydrocephalus have a high degree of success, although there may be complications. Bone marrow transplants for mucopolysaccharidoses are still in the experimental stage, but no other therapy is available. The following table gives some examples of conditions associated with mental handicap, the likelihood of them being recognised at birth, and whether definitive diagnosis is available.

TABLE 1.

Mental handicap	Usually recognised in the neonatal period by clinical or routine screening tests?	
	Yes	No
Invariable and usually severe	Down syndrome (D) Microcephaly, severe (D) de Lange syndrome (C)	Mucopolysaccharidoses (D) Rubinstein-Taybi syndrome (C) Seckel syndrome (C)
Variable and may be absent	Phenylketonuria (D) Hydrocephalus (D) Treacher Collins syndrome (C)	Sex chromosome anomalies (D) Tuberose sclerosis (D) Prader-Willi syndrome (C/D)

(C) = clinical diagnosis only   (D) = definitive diagnose possible

TECHNOLOGIES FOR POSTNATAL DIAGNOSIS

The foundation stone of postnatal diagnosis is the careful clinical examination of every newborn by an experienced doctor, augmented by selected screening tests. Additional tests may be indicated by a family history of some genetic disorder (e.g. muscular dystrophy), by certain antenatal events (e.g. suspected or

proven rubella), or if a child is being investigated for delayed development. The purposes of trying to make a definitive diagnosis are:
1. there may be effective treatment,
2. there may be important genetic implications for the family,
3. parents are not happy with 'cause unknown' for their retarded child, and this tends to prolong feelings of guilt and anger.

TECHNOLOGIES FOR COUNSELLING

The foundation stone here is human conversation. Tests may be necessary to provide information. For example, tuberose sclerosis is often due to a new dominant mutant gene, but an apparently healthy parent may carry the gene in the absence of any skin lesions or neurological disease. A CT brain scan on both parents is necessary to be as sure as possible that their affected child represents a new mutation.

Counselling requires time and knowledge. The initial information that the newborn baby has a serious problem is a devastating experience for parents. They will usually pass through well recognised stages of emotional response, and must be supported along this psychological journey. Until the journey is complete, full understanding and active cooperation from parents are not to be expected.

Five stages can be recognised in this process, and the remarks parents make can help us to determine the stage they have reached. Some parents complete the journey very quickly, others slowly, some never. It is essentially a grief (bereavement) reaction because they have lost the healthy baby they were expecting, or thought they had. Whether they lose the baby or the normality makes little difference. The stages overlap, and progress is not always smooth.
Stage 1
Emotional shock. They do not hear. Long discussions are fruitless. Kindness is helpful. "Nobody told us anything."
Stage 2
Beginning to listen. They hear, but do not understand. Long discussions are still fruitless. "I am glad it is nothing serious."
Stage 3
Denial. They hear and understand, but cannot believe what they hear. "There must be some mistake." There may be demands for more specialist opinions.
Stage 4
Anger and guilt. Now they believe. They get angry with themselves (guilt), with the doctor (threatened litigation) and everybody else. "Why did it happen to *us*?" When the tears flow, the end of the journey is not far off.
Stage 5
"What can we do to help?" Now the unacceptable has been accepted. Personal and family life can be rebuilt.
The journey is shown graphically in Figure 1.

118

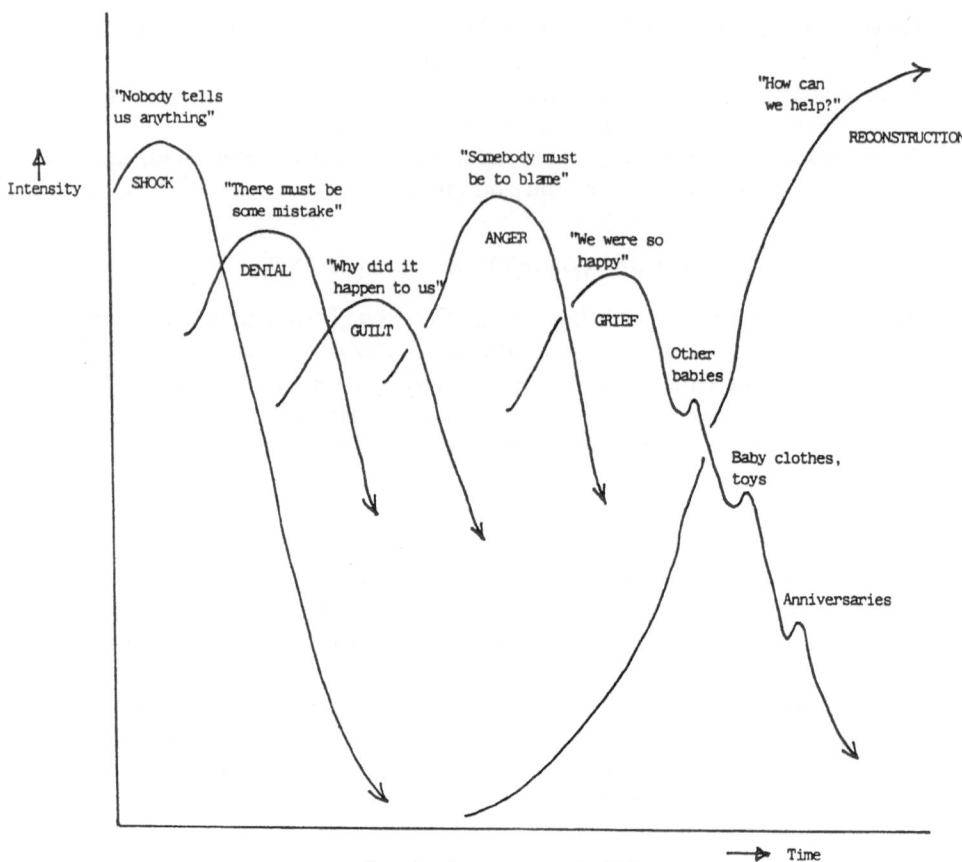

Emotional response to Bad News

Information about the early management of the child; later educational problems; the effect on other children in the family; the effect on personal and sexual relations between parents; genetic counselling: all can be discussed at the right time. But not too much or too soon.

An increasing 'new technology' is in the form of support groups, consisting chiefly of parents of similarly affected children. There are now such groups for a wide variety of conditions.

ETHICAL ISSUES

The ethical aspects of retardation syndromes diagnosed postnatally cannot be separated from those diagnosed prenatally, because:
1) It may be the same syndrome (e.g. Down syndrome).
2) In some countries, termination of pregnancy is legally permitted up to a stage of gestation later than that at which a fetus may survive, so that postnatal

diagnosis in one case could be earlier in gestation than prenatal diagnosis in another.

3) The purposes of pre- and postnatal diagnosis of the same condition may be quite different. For example, research is being carried out for the prenatal diagnosis of cystic fibrosis (not related to mental retardation) so that affected fetuses can be aborted. At the same time, some maternity departments have introduced neonatal screening for cystic fibrosis so that affected infants can be treated early. Thus the attitudes to a 20-week fetus tested *in utero* and a 30-week fetus tested *ex utero* are very different.

In postnatally diagnosed retardation syndromes, and particularly those recognised confidently at birth, the central ethical issue may be seen as the measures which should be taken to prolong life. The most widely debated example is Down syndrome. This can present all three dimensions of the problem. These three, all of which have been practised overtly or covertly, are:

Option 1 Giving 'normal' care, however this may be defined.

Option 2 Witholding normal care, as by giving sedation but no food.

Option 3 Witholding extra care, as by not operating on duodenal atresia.

These in turn give rise to two questions:

1) Is killing (option 2) morally different from allowing to die (option 3)?

2) Is postnatal killing (option 2) morally different from prenatal killing?

If less than maximal care is to be given to the affected individual (and that includes termination of pregnancy), it is necessary to consider the question, 'In whose interest is maximal care being denied?' The only possible answers could be:

1) To save that individual from suffering.

2) To save the rest of the family from suffering.

3) To save society as a whole the inconvenience and expense.

4) To limit the spread of deleterious genes.

In a liberal democracy, reasons (3) and (4) are not acceptable. Reason (1) might be regarded as the only valid one, but in practice it is reason (2) which operates.

Do people with Down syndrome *suffer* from it? Or do they suffer any more than others from rheumatism, infertility, short sight or stomach ulcer, conditions which are not usually punishable by death? I have little experience of adults with Down syndrome. Affected children, apart from the rare case of depression, do not appear to be suffering from their condition.

Mental retardation presents a paradox. The more severe the expected handicap, the more ready will the doctor be to take extreme measures, but the less likely is the affected person to suffer. Severe microcephaly is associated with severe mental retardation but affected children show no evidence of suffering. Indeed, they possibly lack a 'suffering mechanism' except at the level of thalamus and brain stem.

Physically handicapped people of normal intelligence may suffer a good deal, especially at adolescence, but they rarely wish they were dead. On the other hand, they are anxious to do everything possible to ensure that their children are not similarly afflicted.

It is clear, therefore, that any decision to give less than normal care to a mentally handicapped child is made with the object of relieving the family in the first instance, and society in the second, of the demands and inconvenience of a permanently dependent individual. As these decisions are often made soon after the baby's birth, the involvement of the parents, for the reasons described above, can be at best nominal. The doctor is therefore often left to make decisions which ought perhaps to be made by the community he serves.

The burden of caring for a severely handicapped child, especially as such children and their parents get older, is too heavy for many families to bear. The communities in which they live must stop conniving in silence at what the doctors do, and decide whether they are responsible for sharing the care of those who cannot care for themselves, or whether they want the doctors to rid them of the inconvenience.

A. HEIBERG

MENTAL RETARDATION – COMMON DIAGNOSTIC AND GENETIC COUNSELLING PROBLEMS

INTRODUCTION

The heterogeneous aetiology of mental retardation makes diagnosis and genetic counselling difficult for both the families concerned and their physicians. The current definition of genetic counselling includes an information process where those counselled are given facts about the meaning of the diagnosis, aetiology, prognosis and ways to relate to these facts, as well as advice about how to cope with the situation. A precise aetiological diagnosis in young subjects is often uncertain, even after a careful hospital evaluation. The diagnostic process and the quality of medical care are unsatisfactory in most Norwegian institutions for adults with mental retardation. This creates additional burdens for the families when younger relatives ask for counselling, often when already pregnant.

THE FRAMBU HEALTH CENTRE EXPERIENCE

Near Oslo lies a unique information and treatment centre (the Frambu Health Centre) promoting family focused care for disabled children.[2] Families from all over Norway may apply for a two-week stay with all costs, even loss of income for one or both parents, covered by the National Health Insurance. Families may participate in courses on mental retardation in general, or on specialised subjects such as phenylketonuria, Prader-Willi syndrome, fragile X syndrome, Rett syndrome, Laurence-Moon-Bardet- Biedl syndrome, and tuberous sclerosis. There are 37 different courses offered. The information courses provide an opportunity for patients and their families to meet in a setting different from a typical hospital or institution. In addition to the professional staff of Frambu, other specialists also participate. Genetic counselling, rarely available outside Oslo and Bergen, is given by medical geneticists.

About 2/3 of mentally retarded patients coming to Frambu have a correct specific diagnosis such as, for example, Down's syndrome or retardation secondary to meningitis. About half of the other patients have conditions which are aetiologically diagnosable on the basis of such evidence as clinical findings and chromosome or metabolic analyses. In the remaining 1/6 of the total number of cases, some advice on possible recurrence risks may be offered even in the absence of a specific diagnosis. Such advice, based on medical histories and other findings, is often less important for the parents themselves (who may not want further children) than for their healthy offspring and their reproductive pros-

E.K. Hicks and J.M. Berg (eds.), The Genetics of Mental Retardation, 121–124

pects. Experiences shared by parents attending the courses at Frambu may influence reproductive decisions.

SPECTRUM OF MENTAL RETARDATION

The variation between cases is usually wider than parents had visualised and than their counselling physicians may have experienced. Many parents come to realize that other families have more severe problems than do they. Great variation of the degree of mental retardation is seen, even intrafamilialy, between persons affected with such conditions as fragile X syndrome, Prader-Willi syndrome, Laurence-Moon-Bardet-Biedl syndrome, tuberous sclerosis and Down's syndrome. To parents, this variation is difficult to comprehend when they know only a single case. Increased availability of written information in clear understandable language would considerably increase their understanding, particularly if the relevant experience of local physicians and health workers is limited.

The variation in some disorders may be sufficiently great to create doubts about the diagnosis or to raise the question of whether a variant of a particular disorder is present. For instance, one 18-year old boy considered to have Lesch-Nyhan syndrome evidenced severe neurological symptoms and self- mutilation, but wrote essays superior to those of healthy individuals of his age. Although enzyme studies did not show a variant, further DNA studies could be useful in delineating such cases.

The variation may be substantial enough to impede a prognosis of the degree of mental retardation. An affected patient may be within the normal mental range, even when mental retardation is thought to be a cardinal feature of the condition. The Laurence- Moon-Bardet-Biedl syndrome may be an example of this phenomenon. Sometimes mental retardation or clumsiness is a stigmatising diagnosis when visual handicaps of undetected retinitis pigmentosa are present; actually, few such patients seen were outside the normal range (publication in preparation).

RISK OF RECURRENCE

The definition of some syndromes may not only cause diagnostic difficulties, but also impede correct recurrence risk determinations. Smith-Lemli-Opitz syndrome is an example of a multisymptom syndrome with few, if any, constant criteria. Different syndromologists may classify the same patient differently. The original syndrome is described as autosomal recessive in nature, but when using a wider definition, the recurrence risk in siblings is probably lower than 25%. Clinical heterogeneity, also found in other syndromes, makes DNA banking an important consideration when feasible. This would facilitate comparative analysis of well-documented families. No system currently exists which would provide a data base for such future studies.

Recurrence risks are often difficult to understand on the emotional level for parents. When the counsellor is only able to provide risks such as 'either 0 or 25%' or '2–5% on the average,' the situation is obviously worse for prospective parents. If the disability involved shows considerable variation in functional severity from normal to severe mental retardation, then reproductive decisions become even more difficult. Parents often seem more willing to accept a high recurrence risk for a child with a lethal handicap than a low recurrence risk for a child with normal life expectancy and moderate mental retardation.

THE POSSIBILITY OF X-LINKED RECESSIVE VERSUS AUTOSOMAL RECESSIVE DISEASE

When a specific aetiological diagnosis of mental retardation cannot be made in a family with two affected brothers, the question of possible X-linked recessive disease arises. This commonly becomes an important consideration when the parents of the affected boys are deciding about a further pregnancy or when the sister(s) of the boys are contemplating having children. Based upon the relative length of the X chromosome to the autosomes, a rough estimate may suggest that 1:19 multiple congenital anomalies/mental retardation syndromes are X-linked. However, X-linked genes are much more frequent and odds of 1:2 for autosomal recessive to X-linked recessive inheritance have been suggested for non-specific mental retardation.[1] Chromosome and DNA studies of relevant families are needed in the future for the resolution of these dilemmas.

THE NEED FOR FURTHER FAMILY INFORMATION

Important diagnostic points can be ascertained when family information is collected. Fragile X syndrome, neurofibromatosis, myotonic dystrophy and tuberous sclerosis are among many well known examples where this applies. The Norwegian population is very stable and valuable additional data can often be found in old hospital records of relatives. Therefore, the recent trend to destroy hospital records after 10 years is unfortunate and should be discouraged.

Extended family studies are generally possible with the cooperation of family members and physicians. The need for such studies and the ensuing predictive value for identification of asymptomatic carriers is largely unappreciated. More precise risk figures for various relatives in autosomal dominant diseases might be the result of these endeavours.

HOW TO COPE WITH THE SITUATION

Courses for parents of mentally retarded children such as those mentioned earlier are effective for providing information and evaluating services. However, the benefit most appreciated by parents attending a Frambu course was the prospect of meeting other parents in a similar situation and discussing problems and coping strategies with them. Lay organisations can also be helpful in these regards by providing opportunities to meet similarly affected persons and to discuss reproductive attitudes. This is often more advantageous than discussing these issues with physicians who may have only fragmentary knowledge about a specific condition. Contacts with patient groups with rare congenital diseases is also of great educational value for physicians working in the field of chronic disabling diseases.

BIBLIOGRAPHY

1 Herbst, D. S. and Miller, J. R.: 1980, *Nonspecific X-linked mental retardation II: The frequency in British Columbia,* Amer. J. Med. Genet., 7, 461.
2 Storhaug, K. and Vandvik, I. H.: 1983, *Frambu Health Centre: Promoting family focused care for disabled children,* Int. J. Rehab. Res., 6, 175.

G. R. DUNSTAN

TREATING THE SEVERELY HANDICAPPED NEWBORN CHILD: THE FIELD OF CHOICE

Life would be simpler – though less interesting – if duty were always clear. But duty is not always clear. We have to choose, to decide what to do. And ethics, for professional purposes, means moral reasoning directed to decision or choice. A moralist who is not a practitioner is driven to rough-hew his ethics in relation to the extreme case; he will then wish to refine his artefact, in discussion with the practitioner, to match the intermediate cases, the less extreme. Practitioners may find this irksome, because the extreme cases are probably only a minority in daily practice, and because the generalisations on which the moralist has at first to rely do not specify the nature and degree of handicap as a diagnostician would. But the method may help us to display the relevant principles and the logic of their application.

Given, then, a severely handicapped newly-born child, the options are notionally four: to kill; to do nothing at all; to do everything technically possible; or to do what, in informed clinical judgement, is in the child's best interest.

The choice must be made from knowledge, specialist knowledge. It is therefore a professional decision, responsibility for which is laid by society on the clinician in charge. We may note in passing a present-day surge of sentiment which would diffuse that responsibility, or locate it elsewhere, in a committee, for instance, or in parents arbitrarily invested with a "right" to decide. But in the conventional ethics, and in the law, the duty is a professional one. In accepting the post of consultant a practitioner accepts that responsibility. Doubtless he may and should consult – with clinical and nursing colleagues attending the patient, and with parents; but the responsibility for a final decision remains his own. This is clearly expressed in statements of the British Medical Association[2] and the British Paediatric Association.[3] The consultant's decision must be made in accordance with common and accepted principles, grounded in a reasoned morality and upheld by the ethics of his profession and by the law. It is the moralist's business to open out these principles; it is not his business to usurp the duty of decision.

In the simplest analysis the principles or axioms are three. The first is that we place a high value on life, a paramount value on human life. Life is precious. The presumption in medicine is always in favour of life. The presumption begins, some would insist, when life begins, assuming that they know when that is. More defensibly, it is present once the embryo is safely implanted in the womb; though some would extend the presumption backwards to when individuality is estab-

E.K. Hicks and J.M. Berg (eds.), The Genetics of Mental Retardation, 125–131.
© 1988 by Kluwer Academic Publishers.

lished with the primitive streak, even *in vitro*. The presumption grows stronger as the embryo and fetus advance towards maturity.[8] It lasts while life lasts: until, that is, death has been determined by the accepted medical criteria.[12] The presumption in favour of life is rebuttable only on strict grounds sanctioned by morality and law.

The second principle is that we do not, however, place an absolute value on life; we do not assert an absolute right to life – "absolute" meaning admitting of no exceptions. The codes, moral and legal, which protect life – including those of the Judaeo- Christian Scriptures – also allow circumtances when life may legitimately be taken; they include war, capital punishment, strictly defined police activity and self-defence. They admit also that life may be sacrificed voluntarily for sufficient cause. So we can distinguish between justifiable homicide and murder, and between ignoble suicide and noble self-sacrifice.

The third principle is that we do not, nevertheless, allow an absolute or unrestricted right over life. Life may be taken only for certain grave causes, and by persons sufficiently authorized to take it: soldiers, for instance, or the public executioner, or the policeman to whom no less drastic course is available to protect the lives of others or his own. We protect "innocent" life, we say; and innocent commonly means legally or morally blameless. So a man's life may not be taken unless he forfeits it by grave crime proved against him. But "innocent" has another meaning, dating from the time when our ethics and law were worked out in the Latin tongue. It is the negation of the concept in the verb *nocere*, familiar in the maxim *primum non nocere*, "first of all do no harm". "Innocent" in this sense means threatening no hurt or harm. So in war an enemy soldier, morally blameless as he may be, may be shot at while he is armed and threatening harm. But as soon as he lays down his arms or is disabled, he is innocent in the second sense; his life must then be protected.

These are the three basic principles or axioms. Their application passes into the realm of clinical judgement, based on the exactitudes of diagnosis and prognosis. From the first principle, that life is precious, that the presumption is in favour of life, it would follow that the doctor's assumed duty is to preserve, promote and enhance life to the limit of his ability and responsibility, and by proportionate means. So, given a baby with very low birth weight or gestional age, he will apply tentatively, experimentally, the techniques of neonatal intensive care – and take all his readings, meanwhile, to help him to judge whether his attempt, his tentative effort, is going to succeed.

From the second principle, that there is no absolute right to life, it would follow that the doctor is under no strict or absolute duty always to prolong life, irrespective of all other considerations. The duty is a reasonable one, qualified by considerations of proportionate good and harm to the child's interest. He has a duty to discern that interest and to manage accordingly. He may decide, from his scans and tests and from accumulated and recorded clinical experience, that his tentative resort to intensive care will not pay off, and that it both could and should be withdrawn.

From the third principle, that there is no unrestricted right over life, it would follow that the doctor is not at liberty to kill: he is not authorized to do so, as the soldier or others may be in defined roles. His duty is to give care and management appropriate to the patient in his actual condition, and with a view to an outcome either foreseeable or reasonably to be expected.

The criteria for prognosis and management in neonatal intensive care are being tentatively established.[13,19] The criteria by which to decide management for children born with severe handicap have been discussed for a longer period.[5,9,14-16,20-22] In Britain, policy in the 1970s was strongly influenced by a study conducted throughout the 1960s in the Sheffield Children's Hospital. Taking advantage of the ventricular shunt and of antibiotic treatment for infections, all children brought to the Hospital with myelomeningocele were treated aggressively, and the results were observed not only in terms of mortality, survival and morbidity, but also of social consequences for the families concerned. As a result criteria were laid down, and widely accepted, for selective treatment; and a decision was called for soon after birth. At the extremes, infants with mild handicap, atresias and the like, for whom surgical intervention would prove beneficial, were operated upon appropriately. For others, whom surgical intervention would commit to a life of foreseeable *minimum* handicap and suffering, extending perhaps into early adulthood, and requiring meanwhile increasingly aggressive surgery, the management advised was nursing care only, with palliative or sedative treatment, in the expectation that life would be short. This policy received offical encouragement from the Department of Health and Social Security[4] and its results were observed in a changed pattern of mortality statistics. Recent work has modified the criteria and consequent management policy.[17] Ten years (1971-1980) of experience in Newcastle upon Tyne with babies with myelomeningocele and high lumbar paraplegia at birth has demonstrated that, first, the decision whether or not to operate need not be taken within hours of birth, but may be deferred without inevitable adverse effect. Secondly, the survival rate of children not operated upon at birth was longer than had been supposed; and one major factor extending life expectancy was the involvement of parents and family in the child's early care. The children surviving to the age of school-entry and after were chair-bound and incontinent; but none was intellectually retarded, and many were no more handicapped than those given immediate surgery at birth. The demands made on parents were high; and an assesssment of their capacity to meet those demands was necessary as part of the prognosis for the child. Parental involvement helps, rather than hinders, the grieving process should the child die.

There is room for speculation, no doubt, on the differences between parental care and hospital care; but that is a subject for further study. The ethics of the matter now before us turn on the question, How may the child's best interest be served? What means, granted our three axioms, are obligatory, what are elective or optional, and what are forbidden? That the interest of the child is the critical issue has been determined in English and American courts. A child was born in

Hammersmith, England, in 1981 with Down's syndrome and intestinal atresia. The parents refused consent to an operation to remove the blockage which, surgeons said, would give the child a life expectancy of 20 to 30 years. The Court of Appeal, in wardship proceedings, authorized the operation on the ground that it was in the child's best interest that it should be enabled to live: the judge in the inferior court had erred, the Appeal Court stated, because he was influenced by the views of the parents instead of deciding what was in the best interest of the child. (*The Times* Law Report, Re "B" (minor), 8 August 1981). In the U.S.A. case the judgement on the child's interest turned the other way. A child was born in Long Island, New York, with microcephaly, spina bifida and other complications. Surgeons predicted death within two years without surgery, but a life expectancy of severe disablement and retardation for 20 years with it. The surgeons, with parental approval, decided not to operate. A "right to life" group obtained a court order instructing the surgeons to operate. The case went eventually to the U.S. Supreme Court, which upheld the reversal of the lower court's instruction; it was adjudged not to be in the child's best interest that her life should be surgically prolonged, given such a prognosis (*The Times*, 6 and 13 December, 1983). The choice of verbs in the two cases is significant. The English court "authorized" the surgeons to operate, leaving them with their clinical discretion whether to do so or not; the American court "instructed" them. The point is of importance in the standing of the profession and of professional judgement before the law.

These two apparently contradictory decisions, the one authorizing intervention, the other permitting non-intervention, are justified within the terms of a long-standing moral tradition. Since Hippocrates at least it has been recognized that it is physicianly and praiseworthy to realize in what cases medicine is powerless. The Western Catholic tradition has developed the language of "ordinary" and "extraordinary" means of prolonging life.[6,18] The formula is now widely adopted in medical discussion.[10] "Ordinary" means are those which, given the presumption in favour of life and the normal human obligation to cherish it, a phyician is morally bound to offer and a patient is morally bound to accept. "Extraordinary" means carry no such obligation; they remain elective, of choice, though there may be weighty moral indications against their employment, and no moral obloquy attaches to their refusal.

The terms do not refer to the novelty or intricacy of the technology or surgical intervention. They refer to their effect on the patient, each particular patient – whether they would impose on him or his family undue suffering, hardship, distress, disfigurement, impairment of human relations and the like, while offering no cure or remedy for the condition or reasonable hope of benefit. A remedy specific for a particular condition may be ordinary for one patient, extraordinary for another. An antibiotic would be ordinary to combat pneumonia in a young athlete; it would be extraordinary for a patient in whom the pneumonia had supervened upon an advanced terminal cancer.

The choice before a paediatrician given a severely handicapped newly born child may, therefore, be worked out in terms such as these. Given the Newcastle modification of the Sheffield criteria, the selection of infants for surgical intervention, minor or major, and active treatment, or for nursing care only, can be more leisured than was formerly believed. It will take into account not only the physical condition of the child, but also the moral, emotional and other capacities of the parents to involve themselves in his management. Given a favourable prognosis on these counts, active intervention would be an ordinary obligation. Without it, to commit the child to an ever-increasing burden of suffering, without hope of alleviation or remedy, would be extraordinary and not of obligation. Appropriate management would be nursing care.

Nursing care is not a euphemism for abandonment or for the intentional hastening of death. It would include fluid to combat dehydration. Presumptively it would include nutrition: food is so natural a necessity to life that the presumed duty of providing it would be rebutted only by adverse effects in the patient, like renal or intestinal complications, or by disproportionate difficulty in administering it. Analgesic treatment would be in doses appropriate to evident or assumed distress; to give more would be to invite suspicion of a guilty intent.

The distinction between deliberately shortening life and refraining from action to prolong it exercises philosophers more than it troubles practitioners.[11] Omission to act is culpable only if there is a prior obligation so to act. So, a parent who deliberately or negligently left his child to drown in the bath would be as culpable as one who drowned his child in the bath. But, granted the principle that there is no absolute obligation always to prolong life, that necessary condition of culpability is lacking in the present instance. The duty is to provide management appropriate to the patient's condition, doing all that is reasonably necessary in his interest and for his well- being.[1]

Philosophers worry similarly over the principle of "double effect"; though without it moral decision would be impossible in a world where a totally good consequence cannot be guaranteed from even a totally good action. The principle states that an action, good in itself, intended to a good end and effective to that end, is morally licit even though it incurs an unwanted, secondary effect, not intended, but flowing inseparably from it. (It is assumed that no less drastic remedy is available).[7] The principle has wider uses beyond the practice of medicine. A ship's captain in an emergency at sea may be under duty to order the closing of watertight doors in order to save the ship and her company. If it should happen that at the critical time some members of the crew were on the wrong side of the doors, so that as the hold flooded they could not escape, their death would have been occasioned by the captain's order to close the doors. But the captain would not be guilty of those deaths in morals or in law; they occurred as a secondary effect, an inevitable consequence of an action properly taken for a good end. So also in terminal care. Should it happen, as occassionally it might, that measures necessarily taken for the comfort or well-being of a patient had, as an unintended and secondary effect, the hastening of the patient's death, the measures are not therefore culpable, no less drastic remedy being available. The physician is not deterred from using such means by this risk. But the means taken must be evidently proportionate to the end intended and no more.

The moralist, in recognizing the limitation of his own function – not to take or dictate decisions in the medical field which he is not authorized to take – recognizes also the burden of responsibility which he leaves with the doctor. That responsibility is primarily towards each patient as he comes into consideration; but it extends further, into the art or profession of medicine. If the doctor is too hesitant to intervene, to try an untried remedy (*experiri, tentare, attemptare*), he will do little to advance the progress of his discipline; and some children, and their parents, who might have enjoyed a few years of life together will be denied that enjoyment. If he is too zealous, too ruthless even, with innovative surgery or intensive treatment, or presses on with his pet theory without submitting it to the discipline of a research trial or peer review, he may quicken the pace of development in his specialty, or he may not; and his patients may live longer, trying to decide whether or not to thank him.

Assuredly his responsibility is to make the appropriate decisions case by case. In doing so, he is not making technical decisions only. He is a moral agent – the authorized moral agent – making moral decisions, that is, those which subordinate his technique to the common and accepted canons of human good. Medical education should include the forming of practitioners capable of ethical decisions, and increasingly it does. The major problem of the immediate future is to restore to an increasingly consumerist and rights-minded populace the necessary condition of trust.

BIBLIOGRAPHY

1 Anscombe, G. E. and Cusine, D. J.: 1981, *Omission and action*, J. Med. Ethics, 7, 122.
2 British Medical Association: 1983, *Supplementary annual report of council 1982 – 3*, Brit. Med. J., *286*, 1593.
3 British Paediatric Association: 1984, *Care of severely malformed infants*, Arch. Dis. Childh., *59*, 96.
4 Department of Health and Social Security: 1973, *Care of the Child with Spina Bifida*, H.M.S.O., London.
5 Duff, R.S. and Campbell, A. G. M.: 1973, *Moral and ethical dilemmas in the special-care nursery*, New Engl. J. Med.,*289*, 890.
6 Dunstan, G. R.: 1981, *Life prolongation*, In, A. S. Duncan, G. R. Dunstan and R. B. Welbourn, eds., Dictionary of Medical Ethics. Darton, Longman and Todd, London, 266–268.
7 Dunstan, G. R.: 1981, *Double effect*, In, A. S. Duncan, G. R. Dunstan and R. B. Welbourn, eds., Dictionary of Medical Ethics. Darton, Longman and Todd, London, 145.
8 Dunstan, G. R.: 1984, *The moral status of the human embryo; A tradition recalled*, J. Med. Ethics, *10*, 38.
9 Eckstein, H. B., Hatcher, G. and Slater, E.: 1973, *Severely malformed children*, Brit. Med. J., 2, 284.
10 Ethical Working Group: 1985, *Comment on the prognosis for babies with meningomyelocele and high lumbar paraplegia at birth*, Lancet, 2, 996.

131

11 Glover, J.: 1971, *Causing Death and Saving Lives*, Pelican Books, Harmondsworth.
12 Jennett, B.: 1981, *Death, Determination of*, In, A. S. Duncan, G. R. Dunstan and R. B. Welbourn, eds., Dictionary of Medical Ethics, Darton, Longman and Todd, London, 128–130.
13 Kitchen, W. H., Rickards, A. L., Ford, G. W., Ryan, M. M. and Lissenden, J. V.: 1985, *Liveborn infants of 24–28 weeks gestation: Survival and sequelae at two years of age*, In, R. Porter and M. O'Connor, eds., Abortion: Medical Progress and Social Implications. Ciba Foundation Symposium 115, Pitman, London, 122–132.
14 Lorber, J.: 1971, *Results of treatment of myelomeningocele*, Dev. Med. Child Neurol., *15*, 279.
15 Lorber, J.: 1972 *Spina bifida cystica: Results of treatment of 270 consecutive cases with criteria for selection for the future*, Arch. Dis. Childh., *47*, 854.
16 Lorber, J.: 1973, *Early results of selective treatment of spina bifida cystica*, Brit. Med. J., *4*, 201.
17 Menzies, R. G., Parkin, J. M. and Hey, E. N.: 1985, *Prognosis for babies with meningomyelocele and high lumbar paraplegia at birth* Lancet, *2*, 993.
18 Pius XII, Pope: 1957, *Allocution on re-animation*, , Acta Apostolicae Sedis xxxxiv, pp. 1027–1033.
19 Saigal, S., Rosenbaum, P., Stoskopf, B. and Sinclair, J. C.: 1984, *Outcome in infants 501 to 1000 gm birth weight delivered to residents of the McMaster Health Region*, J. Pediat., *105*, 969.
20 Swinyard, C. A., ed.: 1978, *Decision Making and the Defective Newborn*, C. C. Thomas, Springfield, Illinois.
21 Working Party Report: 1975, *Ethics of selective treatment of spina bifida*, Lancet, *1*, 85.
22 Zachary, R. B.: 1968, *Ethical and social aspects of treatment of spina bifida*, Lancet, *2*, 274.

DISCUSSION: POSTNATAL DIAGNOSIS

Computer aid in diagnosis:

Although computers are currently being utilised as a postnatal diagnostic aid, their prenatal use remains minimal in view of limited accuracy. Indeed, it is still difficult to guarantee accuracy at the postnatal stage. To date, computer technology has primarily been utilised to assist in the diagnosis of the large number of relatively rare syndromes, and cannot as yet be a general diagnostic tool.

Artificial insemination:

Although artificial insemination has, for some time, been considered an acceptable alternative in the prevention of certain fetal abnormalities, no appropriate regulatory controls have been established to determine donor qualifications. There is the additional factor that, in many countries, there are fewer ethical problems associated with sperm donation than with ovum donation.

'Best interests' issue: child/parents/society:

What constitute the 'best interests' of the fetus/newborn, and how are these to be determined? Should this be the overriding criterion, or should the parents and the social environment also be taken into consideration and, if so, to what extent?

Involved in the decision-making process in relation to the 'best interests' issue are such factors as:

a. keeping children alive who will suffer severe lifelong handicaps,

b. prenatally diagnosed non-lethal handicaps,

c. the possibility that the prognosis offered in either the prenatal or early postnatal phases may not prove correct,

d. the irreversibility of a decision to terminate pregnancy.

Moreover, attitudes of the parents must be taken into consideration, as it is they who will usually be primarily responsible for the child's support and maintenance. That the best interests of the parents are important in the decision-making process becomes apparent in situations where a professional judgement to perpetuate the life of the fetus must be reconciled with the parents' decision to terminate pregnancy because they cannot, or will not, cope with the burden of a handicapped child.

133

E.K. Hicks and J.M. Berg (eds.), The Genetics of Mental Retardation, 133–135
© 1988 *by Kluwer Academic Publishers.*

On moral-ethical grounds, the best interests of the parents cannot be considered as paramount. Moreover, if the parents are not, for any reason, competent to cope with a handicapped child, other arrangements must be made for its support and maintenance. The only exception to this rule would be where the health of the mother is at risk during delivery. There was general agreement, however, that a constellation of interests exists, and must, as such, be considered. Nevertheless, the overriding principle must remain the best interests of the child. A major difficulty is how these interests should be evaluated and included in the decision-making process.

Termination of pregnancy:

The question of best interests included a discussion concerning pregnancy termination, which in turn involves legal criteria of fetal viability.

Although actual viability is usually much earlier than legal viability, the latter consideration can aid in determining the legal difference between the fetus and the child. There was considerable discussion regarding the wisdom of putting a clinically variable date into a penal statute. This also raised the question of the legal status of a fetus in different countries.

In the U.S., there have been a few cases where the fetus was made a juridical person with legal rights, whereas in English common law the fetus has no actual rights until after live birth (though has 'interests' which the mother is legally bound to protect). In Italy, pregnancy can legally be terminated prior to viability, the physician determining when the latter occurs.

It was generally agreed that, although many societies place moral and/or legal restrictions on pregnancy termination, most countries do not, in fact, offer the necessary financial and institutional support systems to sustain their legal/moral position.

The ethicists present were asked whether it is morally justifiable to apply similar life and death judgements to both the child and the fetus, and, if so, at what specific time *in utero* the fetus becomes a child? Their replies indicated a general agreement that, although moral protection increases with morphological growth, social personhood is usually not assigned until after birth. Moreover, no clear determination has actually been made as to when, and to what degree, society owes protection to the fetus.

Diagnostic uncertainty and pregnancy termination:

Compounding the ethical/moral problems associated with pregnancy termination is the question of diagnostic uncertainty. It was felt by some that, although diagnostic certainty is increasing with new technical developments, there often remains a degree of uncertainty which dictates that the fetus be given 'the benefit of the doubt.' This problem is further compounded when the respective parents are not predisposed to accept any risk, preferring to 'try again.'

The problem of diagnostic certainty initiated a lengthy discussion on the merits and demerits of directive versus non- directive counselling. Particularly where the clinician cannot provide a clear picture of the pregnancy outcome, it was felt that parental autonomy should reign, and that parents must always be informed as clearly as possible about the risks involved and the degree of diagnostic certainty.

Unfortunately, diagnostic uncertainty considerably increases anxiety in parents who must make the decision as to whether or not to terminate a pregnancy. One suggestion was that such uncertainties might better not be made known to parents. Some preferred a directive counselling approach in cases of diagnostic uncertainty, especially where there was a good chance that the fetus was normal. In such cases pregnancy termination could be discouraged.

Other issues related to directive vs. non-directive counselling are summarised below in the discussion section on genetic counselling.

SECTION III

GENETIC COUNSELLING

J. M. BERG

GENETIC COUNSELLING IN MENTAL RETARDATION: REALITIES AND IMPLICATIONS

INTRODUCTION

With the substantial inroads into the understanding of the aetiology of mental retardation during this century, particularly in recent decades, and the consequent dissipation of some quaint mythologies about causation (for example, that mental deficit is a punishment for parental sins, or a kind of degenerative state transmitted by unworthy people), genetic counselling in mental retardation has now attained essentially the same level of sophistication and accuracy as applies, in general, to other defects and disabilities.

With growing frequency, such counselling is being sought concerning the chances of occurrence or recurrence of mental retardation in families and about possible preventive measures. In my experience, virtually all those seeking the counselling belong to one or other of the categories shown in Table 1, the

TABLE 1
Persons who seek genetic counselling concerning mental retardation

a. Parents who have previously had a retarded child
b. Other relatives of such a child
c. Potential parents who are themselves retarded, usually relatively mildly so
d. Potential parents without a family history of retardation who have heard or read of possibly hazardous circumstances applicable to themselves (e.g. advanced maternal age; consanguinity)

majority being in the first of these. Current realities and their implications in the provision of genetic counselling to persons referred to in the table are considered in this paper. The subject has biomedical, psychosocial and ethical ramifications. The latter two are dealt with in other texts in this book, so that the focus here is primarily on biomedical aspects.

APPROACHES

Mental retardation, howsoever defined, is not a unitary, neatly circumscribed entity. It differs not only in its degree of severity, but, even more importantly from the present perspective, in its wide-ranging aetiological heterogeneity. Causal factors of many kinds, of genetic or environmental origin, can result in mental deficit, often accompanied by various physical manifestations (recognisable clinically and/or by special investigations such as biochemical, cytogenetic and radiological ones) which are sufficiently characteristic to constitute discrete syn-

139

E.K. Hicks and J.M. Berg (eds.), The Genetics of Mental Retardation, 139–147
© *1988 by Kluwer Academic Publishers.*

dromes. Many examples of these syndromes are pictorially well illustrated in atlases concerned specifically with mental retardation entities.[5,8]

When dealing with individuals in whose families mental retardation is present, or who are themselves retarded, a precise causal diagnosis is of crucial value for accurate genetic counselling. Such a diagnosis has become increasingly (though by no means always) possible, particularly in instances of relatively severe mental retardation, and is facilitated by examinations of the kinds mentioned above. A detailed family history can also be important in predicting risks of recurrence, as can, in various situations, carrier detection tests. In the absence of an aetiological diagnosis in an affected family member or members, or where concerns about childbearing prospects have arisen in persons without a family history of mental retardation, the genetic counsellor generally has to rely on empiric risk figures as a guide to the chances of mental retardation occurring in the circumstances in question.

There are many nuances (and, indeed, pitfalls) in relation to the considerations mentioned. Some are discussed with reference to various types of mental retardation and their genetic counselling implications in the following sections. Suffice it to add that such counselling is properly concerned not merely with statistical odds of disability occurring, but also with perceptions of potential parents about the burden of the given disability (i.e. how serious it is seen to be) to themselves and to offspring, and with available preventive options (most notably, at present, prenatal ones).

SPECIFIC GENE DEFECTS

Though often individually rare, many syndromes or entities in which mental retardation occurs are due to specific harmful genes with characteristic Mendelian patterns of transmission, generally autosomal recessive, autosomal dominant or X-linked recessive. [9] Unequivocal diagnosis of these conditions in affected individuals provides the basis for mathematically precise calculations of recurrence risks. Thus, for example, the heterozygous parents of a child with typical phenylketonuria (autosomal recessive) have a 25% chance that any other child of theirs would also be affected; a person with tuberous sclerosis (autosomal dominant) of either sex, if fertile, has a 50% chance of having an affected offspring; and a mother carrying the gene for X-linked recessive hydrocephaly, associated with stenosis of the aqueduct of Sylvius, has a 50% chance of producing affected sons and the same chance of producing carrier daughters like herself. In some instances, even though there are no known affected relatives, high risk potential parents can be recognised by suitable carrier detection tests, usually biochemical in nature. This can be a practical approach on a fairly large scale when sections of a population most likely to be the carriers for a given condition can be identified, as, for instance, is the case with the gene for Tay-Sachs disease which has a markedly higher frequency in Ashkenazi Jews than in other population groups of European origin.[11]

There are various problems which can result in serious inaccuracies in the application of the Mendelian ratios referred to above. Among these are the occurrence of phenotypically rather similar conditions with different modes of inheritance (e.g. autosomal recessive Hurler's disease and X-linked recessive Hunter's disease, which can be distinguished enzymatically, and clinically by the presence of corneal clouding only in the former). There is also the possibility of environmentally determined phenocopies simulating Mendelian disorders (e.g., congenital rubella infection can produce effects suggestive of a diagnosis of autosomal recessive Usher's syndrome). Another dilemma can arise in trying to determine whether a condition like tuberous sclerosis, with its widely variable manifestations including cryptic ones, has occurred as a result of new gene mutation (with little risk of a sibling being affected) or by direct transmission of the gene by a parent who carries it (with a consequent 50% recurrence risk in any other child of that parent). Furthermore, a familial incidence of mental retardation in keeping with a Mendelian pattern of transmission does not, of course, necessarily mean that such transmission has occurred. Also, although parental consanguinity is a useful clue in the recognition of rare autosomal recessive disorders, including those associated with mental retardation, a mentally retarded child born to consanguineous parents is not, in itself, proof that those parents face a 25% recurrence risk.

Thus, causal heterogeneity of apparently similar phenotypes, diagnostic difficulties resulting from variability of phenotypic expression of many Mendelian genotypes and suggestive (but inconclusive) family data are all factors which can lead to erroneous conclusions about the presence of specific gene defects. They must be constantly borne in mind in the provision of genetic counselling in relation to such defects.

CHROMOSOMAL ABERRATIONS

As a result of advances in cytogenetic examination methods, such as high resolution banding techniques[14], an increasingly large number of morphologically recognisable chromosome aberrations (and apparently clinically innocuous variants) have been identified since the first demonstration, only 30 years ago, of the normal human diploid chromosome number of 46.[13] Some of the aberrations, generally involving excess or deficiency of chromosome material (both autosomes and sex chromosomes), are lethal *in utero*. Others are observed in live-born individuals with a frequency ranging from relatively high (as is the case with trisomy 21 of Down's syndrome which is generally found in approximately 1 in 700 live-births) to very rare. They are usually associated with mental retardation of different degrees and various concomitant physical manifestations; the mental deficit is most often (though not always) more severe in those with autosomal, than in those with sex chromosomal, aberrations.

The chances of occurrence or recurrence of chromosome abnormalities differ in different situations and can be derived in various ways. Among common relevant factors are advancing maternal age with respect to risks of non-disjunction, the previous delivery of a *de novo* chromosomally abnormal conceptus or live-born infant, or one with an unbalanced karyotype resulting from a parental chromosome peculiarity such as a balanced translocation or certain inversions. In most such circumstances, the actual risk of a future child having a chromosome aberration cannot be calculated with the exactness applicable to conditions with characteristic Mendelian modes of transmission discussed earlier. However, in that case, empirical data often provide good indications of the statistical odds of a particular aberration occurring. Such odds, to take as an illustration the chances of having a live-born Down's syndrome child, range from as little as about 0.05% for chromosomally apparently normal young parents with maternal age around 20 years, to as much as 100%, irrespective of parental age, if one of the parents carries a balanced 21q/21q translocation. A striking example of the latter circumstance was a phenotypically normal 21q/21q translocation carrier mother all of whose 8 liveborn children were reported to have Down's syndrome (her 4 other pregnancies ended in spontaneous abortions at 2 to 3 months of gestation).[4] Risks can differ empirically from the theoretical expectation, as is apparent, for instance, when a parent has a balanced 14q/21q translocation; though it could be expected that one-third of offspring would have Down's syndrome, in reality this syndrome is found in approximately 10% of live-births when the mother is the carrier of the translocation and substantially less frequently when the father is.[12]

An example of a recently extensively documented, and comparatively common, chromosomal finding of a different kind which can be very informative for genetic counselling is what is often called a fragile X (one with a fragile site at Xq27 or Xq28). The fragile X is a heritable marker for a particular type of X-linked mental retardation affecting males and, with less severity and frequency, some heterozygous females.[10]

Detection of these and many other currently recognisable chromosomal anomalies is important in genetic counselling concerning mental retardation, because they are usually accompanied by such retardation and/or increased chances of its recurrence in relatives. The various circumstances connected with increased risks of bearing a child with a chromosome aberration provide, at present, by far the most frequent indications for diagnostic prenatal genetic amniocentesis and, more recently, chorionic villus sampling.

AETIOLOGICALLY UNDIFFERENTIATED MENTAL RETARDATION

In some respects, the aetiology of the mental retardation associated with specific gene defects and chromosomal aberrations discussed above can be regarded as undifferentiated, because the precise causes of these defects and aberrations, and the exact mechanisms whereby they produce mental deficit, are generally not well understood. Here, however, the term 'aetiologically undifferentiated mental retardation' is used for those forms of retardation not known to belong to the specific gene or chromosomal categories considered earlier, nor known to be determined by particular pre-, peri- or post-natal physical causes of environmental origin such as teratogens, birth trauma and post-natal cerebral infections respectively. Though many of these causal factors can result in mental deficit ranging from marked to minimal, there remain large numbers of persons in both the severely (say IQ below 50) and mildly retarded categories in whom such specific factors are not identifiable. Because some considerations bearing on causation and prognosis (and hence on occurrence and recurrence risks) differ in these aetiologically undifferentiated severely and mildly retarded groups, these groups are considered separately below.

A. Severe Mental Retardation

Though severely retarded individuals usually show evidence of organic cerebral pathology, during life and/or at autopsy, which is the neuropathological basis for the mental deficit, this cerebral pathology is currently causally obscure, in the sense referred to above, in about 50% of cases.[1] In families with such aetiologically undifferentiated severely retarded members, chances of recurrence most often have to be inferred from empirical data. Since such data are derived from samples which differ in composition from survey to survey and which are bound also to be causally heterogeneous, it is not surprising that different studies do not always yield similar recurrence risk figures. Nevertheless, a recurrence risk in subsequently born siblings of sporadic cases of aetiologically undifferentiated severe mental retardation ranging from 3 to 6% has been fairly frequently found, with generally lower incidence in more distant relatives of the initial index case. Such factors as the sex of the index case and the presence of some associated neurological manifestations may alter empiric risk figures and data of these kinds are gradually accumulating.[7] Such empirically derived percentages are of some value in genetic counselling, but should be quoted in the context of an explanation of how they were derived as an indication that they may not be valid in the particular case under consideration because unknown relevant factors may be operative in that instance.

A number of clues in individual families in which aetiologically undifferentiated severe mental retardation has occurred can provide a firmer basis for estimating recurrence risks in those families. For instance, a characteristic familial distribution if several relatives are affected may make one or other Mendelian pattern of transmission quite likely even though a specific gene defect has not been identified. Another example of a circumstance which can add some precision to recurrence risk prediction is the recognition of a distinct mental retardation syndrome, albeit aetiologically obscure, if empirical data on chances of recurrence of that syndrome are available – de Lange syndrome, for instance, which is usually sporadic, has been found to recur in abour 4% of subsequently born siblings[2], and a rather similar percentage has been noted in various countries for the risk of a neural tube defect (often associated with mental retardation) in a first-degree relative of someone who has such a defect.

These approaches, imperfect though they are, are the best currently available for genetic counselling purposes in the circumstances described. Much better evidence of risk size could, of course, result from aetiological clarification of presently obscure forms of mental retardation and one can reasonably anticipate further inroads in that sphere in the near future.

It is necessary to refer here also to a concern frequently raised by parents of a severely retarded individual about that individual's reproductive prospects, particularly with respect to risks that she or he would produce retarded children. That risk largely depends on the cause of the retardation in the affected possible parent. However, whether the cause is identifiable or not, many severely retarded persons are infertile or sterile and thus biologically (and often for psychosocial and clinical reasons also) unlikely to become parents. That in itself may not resolve understandable parental anxieties about a severely retarded daughter or son having children, so that the question of contraceptive precautions or the more sensitive and controversial issue of sterilisation may arise. Space does not permit delving into that issue here. However, the personal conviction is expressed that decisions in any particular instance should be based on evaluation of the relevant biomedical and psychosocial circumstances in that case and not on preconceived notions leading to uniform conclusions in all instances.

B. Mild Mental Retardation

Unlike persons with severe retardation, many who are mildly retarded, without evidence of a specific gene defect, a chromosomal aberration or a physical environmental cause for their disability, do not have recognisable organic cerebral pathology. This has led to a widely held view that a large proportion of such individuals constitute part of the normal, non-pathological distribution of intelligence in the general population and that they differ from those of average intelligence because of the operation of the same kind of factors that result in the latter differing from persons of superior mental ability. These factors are generally considered to be both genetic and environmental (i.e. multifactorial) in the form of beneficial or adverse multiple (poly-) genes and psychosocial influences. There is continuing debate, often heated because of ideological over-tones, as to the relative importance of the genetic and environmental components in this multifactorial model.

In the genetic counselling context, growing emphasis on as normal a life as possible in the community for the mentally retarded has led increasingly to information being sought about child-bearing implications for them, especially for those who are mildly retarded and generally fertile. In the absence of recognis-able specific biomedical hazards of the kinds referred to above, mildly retarded persons are not prone to have children with severe mental deficit and many of these children are normal or near normal in their mental development, particu-larly if not exposed to undue psychosocial disadvantages. In general, whether it is prudent or not for any couple, one or both of whom have aetiologically undifferentiated mild mental retardation, to become parents depends more on psychosocial considerations (for example, the capacity of the couple to cope adequately with the responsibilities of parenthood, and support available to them with regard to their well-being and that of their children) than it does on biological ones. Even if avoiding childbearing in particular instances seems sensible, such a conclusion is not in itself ground for proscribing marriage, or, indeed, sexual relationships outside of wedlock. Whatever moral convictions may be held by anyone in the latter regard, there is no good reason to apply them differently to retarded persons than to normal ones. These broader issues often arise, and deserve sensitive attention, in genetic counselling contexts, especially those concerning mild mental retardation.

OTHER CONSIDERATIONS

Though the main focus above has been on the statistical chances of mental retardation occurring or recurring in various circumstances, it is evident that related considerations are also an integral component of genetic counselling. Some of these are briefly discussed below.

Decisions of parents or potential parents about having children are, of course, dependent on many variables, among which the mathematical odds mentioned are but one. Where there is an increased risk of some form of mental retardation occurring, it is essential to convey information not only on the magnitude of the risk, but also, as comprehensively as possible, on the range or anticipated degree of severity, likely or possible concomitant abnormalities; therapeutic prospects, and prognosis in terms of functional capacities and life expectancy. Even when there is an increased risk of a reasonably well-defined mental retardation syndrome or disease entity occurring, variability in manifestations can make firm predictions about severity and clinical outcome in affected persons unreliable. A male child with an extra Y chromosome, for instance, may or may not develop normally in terms of intellect and behaviour[3], and any child with tuberous sclerosis may show hardly any adverse effects or be gravely abnormal mentally and/or physically.[6] Such variability must be explained to at-risk potential parents as it could well be a significant consideration in their child-bearing decisions. In particular, where there is a previously affected offspring or other family member with a condition associated with mental retardation, it is essential to indicate where applicable (as it usually is) that another affected individual may not be afflicted to the same degree or extent. This type of information facilitates informed decisions.

In addition, where there is an indication for feasible and available prenatal diagnostic investigations, they require to be explained with respect to their nature, rationale, advantages and disadvantages. These procedures, including ultrasonography, amniocentesis, chorionic villus sampling and fetoscopy, are highly relevant for the intrauterine detection of a substantial number and range of abnormal conditions in the fetus which are connected with mental retardation

146

post-natally. As, at present, 'curative' medical or surgical intervention when these conditions are detected is seldom possible then or later, a termination of pregnancy option in the context of prenatally detected abnormality is usually the rationale for undertaking diagnosis. Thus, issues involved in the prenatal recognition of abnormalities linked with mental retardation are not only technical ones (such as safety of the procedures, reliability of results and sometimes problems in interpreting them), but also emotional, ethical and, indeed, legal ones. Prenatal diagnostic considerations and their ramifications are a major component of genetic counselling concerning mental retardation, but are referred to only briefly here because they are the focus of other contributions in this book. The subject will become even more important as new prenatal investigative techniques (for example, gene probes involving DNA technology) extend diagnostic scope – with, as can be expected, future advances in effective therapy as an obviously more desirable alternative to pregnancy termination.

This presentation explicitly maintains, and it is presumably a hardly controversial view, that all who seek genetic counselling are entitled to as comprehensive information as possible concerning their risk of disability in offspring, the expectations and prognosis for those who may be affected, and intervention options. Implicit in the observations made is that decisions about having children, or about feasible pregnancy termination in the presence of fetal abnormality, are essentially the prerogative of the potential parents and not of the genetic counsellor. In general, the latter fulfills his or her role best by being supportive rather than directive concerning informed parental pregnancy decisions.

BIBLIOGRAPHY

1 Berg, J. M., Clarke, A. M. and Clarke, A. D. B.: 1985, *The changing outlook,* In, A. M. Clarke, A. D. B. Clarke, and J. M. Berg, eds., Mental Deficiency: The Changing Outlook, 4th ed. Chapter 1, Methuen, London.
2 Berg. J. M., McCreary, B. D., Ridler, M. A. C. and Smith, G. F.: 1970, *The de Lange Syndrome,* Pergamon Press, Oxford.
3 Brøgger, A.: 1985, *XYY and its relation to criminality,* In, A. A. Sandberg, ed., The Y Chromosome: Part B, Clinical Aspects of Y Chromosome Abnormalities, Chapter 19, Alan R. Liss Inc., New York.
4 Furbetta, M., Falorni, A., Antignani, P. and Cao, A.: 1973, *Sibship (21q21q) translocation Down's syndrome with maternal transmission,* J. Med. Genet., *10,* 371.
5 Gellis, S. S. and Feingold, M.: 1968, *Atlas of Mental Retardation Syndromes: Visual Diagnosis of Facies and Physical Findings,* U. S. Dept. of Health, Education and Welfare, Washington, D. C.
6 Gomez, M. R., ed.:1979, *Tuberous Sclerosis,* Raven Press, New York.
7 Herbst, D. S. and Baird, P. A.: 1982, *Sib risks for nonspecific mental retardation in British Columbia,* Amer. J. Med. Genet., *13,* 197.
8 Holmes, L. B., Moser, H. W., Halldórsson, S. Mack, C., Pant, S. S. and Matzilevich, B.: 1972, *Mental Retardation: An Atlas of Diseases with Associated Physical Abnormalities,* MacMillan Co., New York.
9 McKusick, V. A.: 1986, *Mendelian Inheritance in Man: Catalogs of Autosomal Dominant, Autosomal Recessive, and X-linked Phenotypes,* 7th ed. The Johns Hopkins University Press, Baltimore.
10 Opitz, J. M., ed.: 1986, *X-linked Mental Retardation 2,* Amer. J. Med. Genet., vol. 23, no. 1/2 (Special Issue).

11 Petersen, G. M., Rotter, J. I., Cantor, R. M. Field, L. L., Greenwald, S., Lim, J. S. T., Roy, C., Schoenfeld, V., Lowden, J. A. and Kaback, M. M.: 1983, *The Tay-Sachs disease gene in North American Jewish populations: geographic variations and origin,* Amer. J. Hum. Genet., *35,* 1258.
12 Therman, E.: 1986, *Human Chromosomes: Structure, Behavior, Effects,* 2nd. ed. Chapter 24, Springer-Verlag, Berlin.
13 Tjio, J. H. and Levan, A.: 1956, *The chromosome number of man,* Hereditas (Lund), *42,* 1.
14 Yunis, J. J.: 1981, *Mid-prophase human chromosomes. The attainment of 2000 bands,* Hum. Genet., *56,* 293.

J. P. FRYNS

X-LINKED MENTAL RETARDATION AND FRAGILE X (q27): PITFALLS AND DIFFICULTIES IN DIAGNOSIS AND GENETIC COUNSELLING

At the present time, the fragile X syndrome has become the subject of increasing and widespread medical interest.[1-6,8,11,13,15-19,22,23,25] Recently, a second International Workshop on the fragile X and X-linked mental retardation was held in Australia and resulted in a 737 page special issue of the American Journal of Medical Genetics on the subject.[20] The fragile X syndrome has generally been recognised as one of the major causes of mental retardation in all populations and ethnic groups.[26] Next to trisomy 21, the fragile X syndrome is the most common specific cause of mental retardation among mentally retarded boys. Recent epidemiological studies indicate that the incidence of the syndrome in children of school age is 1 in 1,360–1,500 for boys and 1 in 2,073 for girls.[12,28] Up to now, we have detected fragile X syndrome in 140 index patients. The present paper summarises the different problems we have experienced in the evaluation of patients and their families.

THE PHENOTYPE OF THE FRAGILE X MALE

In postpubertal males, the fragile X syndrome is associated with a characteristic clinical triad: moderate mental retardation, long face and macroorchidism. Systematic fragile X screening studies of mentally retarded male populations have demonstrated, however, that the typical triad is present in only $\leqslant 60\%$ of fragile X positive adult males.[8,14,17,19] In 25% of the patients macroorchidism is absent and in 10 to 15% mental retardation is the only manifestation. Moreover, great intrafamilial phenotypic variability has been found in most of the pedigrees. These data illustrate that clinical selection criteria for fragile X screening in mentally retarded adult males are difficult, if not impossible. In addition, an increasing number of so-called 'typical' fragile X males are known in whom fragile X screening was negative on different occasions using different techniques. Recently, we observed 3 males with acquired lesions of the central nervous system (1 with a tumour of the third ventricle, and the other 2 with a history of peri- and postnatal accidents). In all 3 patients macroorchidism was associated with facial characteristics similar to those found in fragile X males, but they were fragile X negative. The finding of megalotestes associated with a fragile X phenotype in these patients and, more particularly, the documentation of a hypothalamic tumour in the first patient, indicates that acquired hypo-

149

E.K. Hicks and J.M. Berg (eds.), The Genetics of Mental Retardation, 149–156
© 1988 by Kluwer Academic Publishers.

thalamic lesions may result in phenotypic changes clinically indistinguishable from the fragile X syndrome.[9]

In prepubertal males, clinical signs are even more non-specific, and in only 40–50% of young fragile X males moderate mental retardation is associated with facial abnormalities, of which relative macrocophaly is more striking than a long face and large everted ears. Macrogenitosomia is a relatively rare finding ($\pm$ 10%) in the prepubertal males. More characteristic is an apparently typical psychological profile, regardless of the IQ level, with severe hyperkinetic behaviour, emotional instability, hypersensitivity, handbiting and autistic features. In Leuven, an ongoing evaluation of the psychological profiles in fragile X boys and boys with so-called idiopathic developmental retardation shows relevant quantitative differences. Fragile X preschool boys present, in general, mild to moderate mental retardation with severe delay in speech development. Impulsive, hyperactive behaviour problems are more frequent and more intense than in control groups. Autistic features are most pronounced in the group of fragile X boys with IQs below 55.

THE FEMALE AND THE FRAGILE X

Our investigations on female carriers not only concerned the cytogenetic difficulties related to the fragile X expression in the female, but increasingly uncovered the consequences for the mental status and the phenotype of female fragile X carriers.[9] In this study of 144 female heterozygotes, 46 (32%) had subnormal intelligence. An IQ between 85 and 70 was present in 26 cases and most of these had more or less serious problems in the management of their family tasks, with difficulties in raising their children. In 5 of them regular psychiatric admissions were necessary because of chronic psychotic problems. The IQ was between 70 and 55 in 11 carriers; 9 others were markedly mentally retarded.

Partial clinical manifestations in the carriers were common (28%), and facial features included high, broad forehead, long face, and mandibular prognathism. Partial phenotypic expression was more common in the mentally subnormal females (55%), but was not rare in the mentally normal carriers (14%).

Fragile X testing in carriers is disappointing. In the normal female heterozygote repeated tests for the fragile X remain negative in more than 50% of cases. The limitations of the cytogenetic techniques in fragile X carrier detection create great difficulties for accurate genetic counselling of the individual female at risk. Whereas fragile X expression is influenced somewhat by age and mental level of the carrier, the most important factor is the phenotype. In all females with partial phenotypic expression, the fragile X was present regardless of age and mental level. Careful clinical examination of female relatives of fragile X males provides valuable and definite information with regard to the carrier status in individual women.

The X-inactivation pattern in fragile X carriers also is a matter of discussion.[3,8,24] Uchida and Joyce[24] reported the fragile X to be active more frequently than the normal X in mentally subnormal heterozygote females, with the opposite in the mentally normal, but other studies have not confirmed this.[3,8] Our data indicate that BrdU inactivation studies in lymphocytes do not make possible the prediction of the mental status of female heterozygote carriers.

The partial (or complete) clinical manifestation of the fragile X syndrome in the female heterozygote makes recognition of 'X-linked inheritance' in an individual family much more difficult. In more than 20% of the 140 families we have studied to date 'non-specific familial mental retardation' was present. In these families, borderline to mentally retarded mothers (IQ varying between 49 and 70) gave birth to several mentally retarded children, both boys and girls. In these families also, the presence of mental subdevelopment in the mother and one or several daughters failed to indicate a true pattern of X-linked inheritance. The heterozygous fragile X mothers with low intelligence showed both the biological 50% risk of transmission of the fragile X to their progeny, and evidence of a low psychosocial level of care towards their children. This results in high infant death, battered child syndrome, and 'discharge' from parenthood. Fragile X diagnosis in such families is extremely important and is the basis on which help and prevention can be organised for them.

TRANSMISSION OF FRAGILE X THROUGH NORMAL MALES

Male transmission of the fragile X syndrome has been well documented. In the original family described by Martin and Bell in 1943, 2 unaffected brothers had passed on the gene for the fragile X syndrome through their healthy daughters to the next generations in which 11 mentally retarded males were observed. The phenomenon of non-manifesting carriers transmitting the gene for X-linked mental retardation and fragile X has since then been repeatedly observed, and is an apparently frequent condition.[7,10,27]

In the 140 families studied in Leuven referred to above, evidence for transmission of the fragile X chromosome through normal male(s) was present in at least 15 cases. Figure 1 illustrates such a pedigree. Fragile X syndrome was found in 2 boys (IV, 6 and IV, 10); their mothers were sisters (III, 4 and III, 7) and further examination of the family confirmed the diagnosis of fragile X syndrome in 2 of their paternal nephews (III, 9 and III, 14) and 1 paternal niece (III, 13). How puzzling pedigree findings in the fragile X syndrome may be is further illustrated by the finding of a definite case of the syndrome in a boy (V, 1) in the maternal side of the family one year later.

In a number of other families (Figure 2), the occurrence of fragile X syndrome in several maternal nephews (Figure 3), together with true macrocephaly in the maternal grandfather (Figure 4) with negative fragile X screening, may also be an indication of probable male transmission. In these types of pedigrees the normal transmitting males do not show the marker (negative fragile X screening), but transmit the trait to their daughters. In those females, the fragile X site is

152

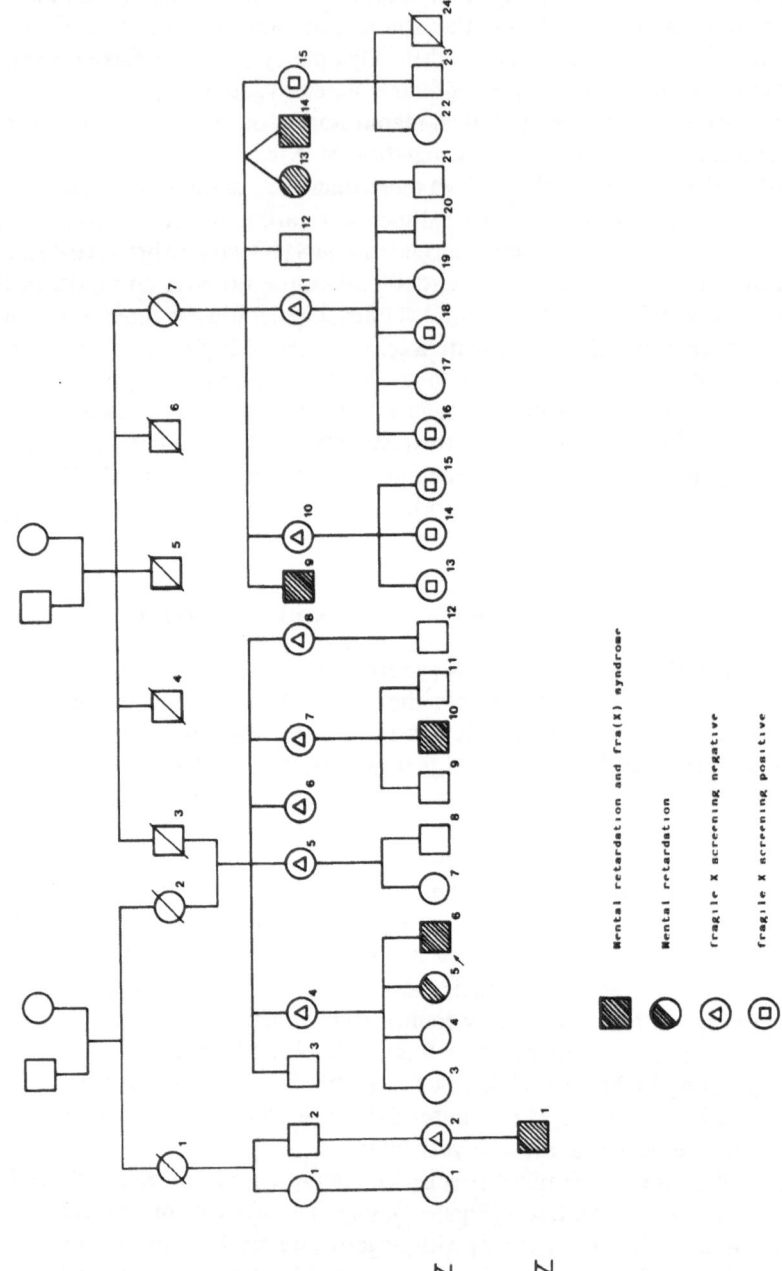

Fig. 1 Pedigree 1

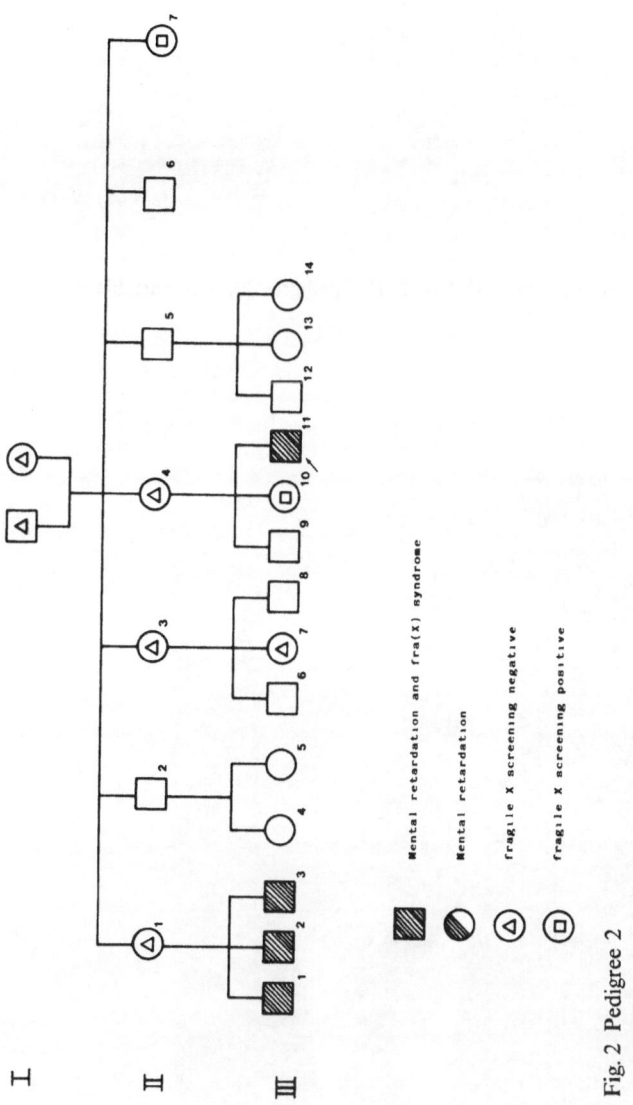

Fig. 2 Pedigree 2

Fig. 3 Three fragile X positive boys (III, 1, 2, II – pedigree 2), with variable facial features

Fig. 4 The maternal grandfather (I, 1 – pedigree 2) with macrocephaly and relatively long face

difficult to demonstrate; in most of them fragile X screening is negative, and, in a few, we found extremely low percentages of fragile X positive cells. These daughters, as obligate heterozygotes, are at great risk of transmitting the marker to their sons. They again can be mentally retarded and fragile X positive.

Available studies do not allow a precise estimate of the ratio of mentally normal versus subnormal males with positive fragile X. Based upon data in prepubertal boys, at least 10–15% of fragile X males may have a sufficiently high level of mental performance to make reproduction socially possible and acceptable.

PRENATAL DIAGNOSIS

The high incidence of mental subnormality in female offspring of heterozygote carriers creates a difficult problem in genetic counselling. It indicates that prenatal diagnosis of the fragile X male does not remove the risk for the female heterozygote to have mentally retarded offspring.[21,29] There is an insufficient correlation between the percentage of fragile X cells, the inactivation pattern of the fragile X, and the mental status in the individual female to solve this problem at present.

Technical difficulties in demonstrating the fragile X in amniotic cells or in cultures of chorionic villi are important, and failures have been reported resulting from low yields of marker X positive cell in amniocyte cultures. Negative and low frequency positive results should be regarded with caution. Further studies are clearly needed, to include fetal blood cultures and the use of restriction fragment length polymorphisms, to complement both pre- and postnatal fragile X diagnosis. This is especially important when the mutation is strongly suspected, though lymphocytes will probably turn out to be a more reliable source with regard to expression of the fragile X site. There are now several probes available for the genetic analysis of the fragile X syndrome but so far none of these is sufficiently closely linked to be of clinical use in pre- or postnatal diagnosis.

BIBLIOGRAPHY

1 Blomquist, H. K:son, Gustavson, K.-H., Holmgren, G. Nordenson, I. and Pålsson-Stråe: 1983, *Fragile X syndrome in mildly mentally retarded children in a Northern Swedish county. A prevalence study*, Clin. Genet., *24*, 393.
2 Blomquist, H. K:son, Gustavson, K.-H., Holmgren, G., Nordenson, I. and Sweins, A.: 1982, *Fragile site X chromosomes and X-linked mental retardation in severely retarded boys in a Northern Swedish county. A prevalence study*, Clin. Genet., *21*, 209.
3 Brøndum-Nielsen, K., Dyggve, H. V., Knudsen, H. and Olsen, J.: 1983, *A chromosomal survey of an institution for mentally retarded*, Dan. Med. Bull., *30*, 5.
4 Carpenter, N. J.: 1983, *The fragile X chromosome and its clinical manifestation*, In, A. A. Sandberg, ed., Cytogenetics of the Mammalian X Chromosome, Part B, Alan R. Liss, New York, 399–414.
5 Carpenter, N. J., Lechtman, L. G. and Say, B.: 1982, *Fragile X-linked mental retardation. A survey of 65 patients with mental retardation of unknown origin*, Amer. J. Dis. Child., *136*, 392.

156

6 Fishburn, J., Turner, G., Daniel, A. and Brookwell, R.: 1983, *The diagnosis and frequency of X-linked conditions in a cohort of moderately retarded males with affected brothers*, Amer. J. Med. Genet. *14*, 713.
7 Froster-Iskenius, U., Bödeker, K. Oepen, T., Matthes, R., Piper, U. and Schwinger, E.: 1986, *Folic acid treatment in males and females with fragile-(X)-syndrome*, Amer. J. Med. Genet., (Special Issue), *23*, No. 1/2, 273.
8 Fryns, J. P.: 1984, *The fragile X syndrome. A study of 83 families*, Clin. Genet., *26*, 497.
9 Fryns, J. P.: 1986, *The female and the fragile X*, Amer. J. Med. Genet., (Special Issue), *23*, No. 1/2, 157.
10 Fryns, J. P. and van den Berghe, H.: 1982, *Transmission of fragile (X) (q27) from normal male(s)*, Hum. Genet., *61*, 262.
11 Glover, T.: 1983, *The fragile X chromosome: Factors influencing its expression in vitro*, In, A. A. Sandberg, ed., Cytogenetics of the Mammalian Chromosome, Part B, Alan R. Liss, New York, 415–430.
12 Gustavson, K.-H., Blomquist, H. K:son and Holmgren, G.: 1986, *Prevalence of the fragile-X syndrome in mentally retarded boys in a Swedish county*, Amer. J. Med. Genet., (Special Issue), *23*, No. 1/2, 581.
13 Herbst, D. S.: 1980, *Nonspecific X-linked mental retardation. I: A review with information from 24 new families*, Amer. J. Med. Genet., 7, 443.
14 Herbst, D. S., Dunn, H. G., Dill, F. J., Kalousek, D. K. and Krywaniuk, L. W.: 1981, *Further delineation of X-linked mental retardation*, Hum. Genet., *58*, 366.
15 Herbst, D. S. and Miller, J. R.: 1980, *Nonspecific X-linked mental retardation. II. The frequency in British Columbia*, Amer. J. Med. Genet., 7, 461.
16 Howard-Peebles, P. N.: 1983, *Conditions affecting fragile X chromosome structure in vitro*, In, A. A. Sandberg, ed., Cytogenetics of the Mammalian X Chromosome. Part B, Alan R. Liss, New York, 431–443.
17 Jacobs, P. A., Glover, T. W., Mayer, M., Fox, P., Gerrard, J. W., Dunn, H. G. and Herbst, D. S. 1980, *X-linked mental retardation: A study of 7 families*, Amer. J. Med. Genet., 7, 471.
18 Jacobs, P. A., Mayer, M., Matsuura, J., Rhoads, F. and Yee, S. C.: 1983, *A cytogenetic study of a population of mentally retarded males with special reference to the marker (X) syndrome*, Hum. Genet., *63*, 139.
19 Mattei, J. F., Mattei, M. G., Aumeras, C., Auger, M. and Firaud, F.: 1981, *X-linked mental retardation with the fragile X. A study of 15 families*, Hum. Genet., *59*, 281.
20 Opitz, J. M., Reynolds, J. F., Spano, L. M., eds.: 1986, *X-linked Mental Retardation, 2*, Amer. J. Med. Genet., (Special Issue), *23*, No. 1/2.
21 Schmidt, A. E., Passarge, E., Seemanovà, E. and Macek, M.: 1982, *Prenatal detection of a fetus hemizygous for the fragile X-chromosome*, Hum. Genet., *62*, 285.
22 Sutherland, G.: 1979, *Heritable fragile sites on human chromosomes. III. Detection of fra(X) (q27) in males with X-linked mental retardation and in their female realtives*, Hum. Genet., *53*, 23.
23 Sutherland, G. R. and Ashforth, P. L. C.: 1979, *X-linked mental retardation with macroorchidism and the fragile site at Xq27 or 28*, Hum. Genet., *48*, 117.
24 Uchida, I. A. and Joyce, E. M.: 1982, *Activity of the fragile X in heterozygous carriers*, Amer. J. Hum. Genet., *34*, 286.
25 Venter, P. A. and Op't Hof, J.: 1982, *Cytogenetic abnormalities including the marker X chromosome in patients with severe mental retardation*, S. Afr. Med. J., *62*, 947.
26 Venter, P. A., Op't Hof, J. and Coetzee, D. J.: 1986, *The Martin-Bell syndrome in South Africa*, Amer. J. Med. Genet., (Special Issue), *23*, No.1/2, 597.
27 Webb, G. C., Rogers, I. G., Pitt, D. B., Halliday, J. and Theobald, T.: 1981, *Transmission of fragile (X) (q27) site from a male*, Lancet, *1*, 1231.
28 Webb, T. P., Bundey, S. E., Thake, A. I. and Todd, J.: 1986, *Population incidence and segregation ratios in the Martin-Bell syndrome*, Amer. J. Med. Genet., (Special Issue), *23*, No. 1/2, 573.
29 Webb, T., Gosden, C. M., Rodeck, C. H., Hamill, M. A. and Eason, P. E.: 1983, *Prenatal diagnosis of X-linked mental retardation with fragile (X) using fetoscopy and fetal blood sampling*, Prenat. Diag., *3*, 131.

A. DUPONT

MENTALLY RETARDED CHILDREN ADMITTED TO A CHILD PSYCHIATRIC DEPARTMENT: DIAGNOSTIC PROBLEMS AND COUNSELLING

INTRODUCTION

The passage of a handicapped child through a health system varies greatly. From the study of a certain group of mentally retarded children (i.e., all those referred to the Danish national care system[3]), it is obvious that this variation is not only caused by differences in the individual child's symptoms and degree of handicapping conditions, but is also very much a result of haphazard events consequent on the reactions of the persons involved. These persons are primarily the parents, but also the doctors of the family, the referral pattern of the doctors to the hospital system (which includes paediatricians, neurologists, and child psychiatrists), and psychologists of the preschool agencies. Also influential is the structure of the health care system itself. A growing tendency to normalise and integrate the handicapped child into the normal system[5,9] has introduced new pathways through it.

In some cases it is obvious, already at birth, that a child is abnormal and will be mentally retarded (e.g., Down syndrome, other chromosomal disorders, some instances of early encephalopathy with different types of epilepsy). In other cases, the family considers the child as normal until some specific retardation of 'milestones' becomes apparent. Early risk children programmes and other medical screening procedures may reveal the retardation in this latter group at an earlier stage. This was stressed by Swedish investigators who also described patterns for the main diagnostic pathways.[8] However, it is surprising that even where there are university neurological and paediatric services plus special facilities for such handicaps as speech and language deficiencies, and hearing and vision problems, children pass undiagnosed as to aetiology to child psychiatric departments, and often do not get a mental retardation diagnosis until they have been seen by a specialist in mental retardation syndromes.

MATERIAL AND METHODS

As advisory medical assistant in the child psychiatric hospital of Aarhus, it was possible for me to examine 20 children after the implementation of a new law on integration in 1980. The children were referred because of such features as psychotic symptoms, behavioural disturbances, and psychomotor retardation. For the entire country (1982), per 100,000, 69 children under 15 years were admitted to a child psychiatric unit (total: 390 children). The Aarhus unit covers

157

E.K. Hicks and J.M. Berg (eds.), The Genetics of Mental Retardation, 157–165
© 1988 *by Kluwer Academic Publishers.*

approximately 1/4 of the country with about 120 admissions per year. Approximately half the cases are first admissions. According to the diagnostic listing, mental retardation constitutes a very small proportion when calculated as 'number diagnosed at first admission', which is only a few per year. However, for all admissions of both sexes, if all additional diagnoses are counted, the number increases to about 15–20 per annum (about 10%) of everyone classified as ICD 310 to 315.

RESULTS

The 20 children in this study, 15 boys and 5 girls, were all first referrals. The main diagnoses were: Psychoses – 4; Perturbatio mentis infantilis non-psychotica – 10; Moderate, or mild mental retardation – 2; Non-specific non-psychotica perturbatio – 4. Mental retardation as an additional diagnosis was used in 13 cases: Borderline – 2; Mild – 1; Moderate – 6; Severe – 2; Unspecified – 2.

Thus, 5 cases had no mental retardation diagnosis, either as a main or as an additional diagnosis. However, in all 20 cases intellectual function was below the age average, according to tests and descriptions either prior to or during hospitalisation.

The average age of the children when examined for mental retardation syndromes was 7 3/4 years (range = 2 1/2 – 17 years); when the 3 children aged over 16 years are excluded, the average age was 6 years. The average age when first referred to the child psychiatrists was 7 years and 10 days (range 1 3/4 – 16 1/12 years). Thirteen of the children had previously been referred to paediatric departments. Fifteen had been neurologically examined by a specialist. Seven had epileptic fits prior to admission.

Examinations undertaken were very extensive. They included:
1. Electroencephalography in 19 cases. Thirteen showed abnormalities, of which 2 were severe.
2. CT scanning in 15 cases. Abnormalities were found in 5.
3. Cytogenetic investigations in 19 cases (in 6 of these for fragile X).[12] In 1 boy with EEC syndrome (ectrodactyly, ectodermal dysplasia and facial clefting) an inversion was found similar to that noted in his normal father. All other chromosomal examinations were negative.
4. Virus studies in 4 cases in which the histories suggested possibile causality. All were negative.
5. Growth data collection related to age from the history, medical records and actual measurements. It was possible to obtain a description of birth for all of the probands (see Table 1). The accuracy of this information was questionable in only 2 cases (one child had been adopted and the other was a refugee). Although the birth weight was below normal (2,500 g) in 2 cases, the gestational age was within the normal limit (38–42 weeks). One child exceeded the normal birth weight (4,500 g), but also had a gestational age in excess of 42 weeks. In 2 cases, dwarfism (height below the average by 3

standard deviations) was diagnosed. One of these had Seckel's syndrome, and the other had an unknown prenatal aetiology. In the latter case all the laboratory tests for endocrine abnormality, prior to parental interruption of examination, were negative. In 7 cases growth was mildly retarded, and in 2 others (both boys) growth at puberty was above normal. Head circumferences were measured and compared with Danish standards.[11] Three cases were microcephalic, 2 had a circumference 1–2 standard deviations below the mean, 1 was macrocephalic, and the remaining 15 were within the normal range (although the shape was abnormal in 2 of them).

6. Analyses for lead, cadmium and mercury in 4 cases with a suspected history of exposure. One of these showed severe lead intoxication.
7. Special examinations by otologists, ophthalmologists and orthopaedists, where indicated. Congenital malformations of the ears were found in 2 cases. One boy with pseudohermaphroditism had severe abnormalities of the external ear on one side, and another boy with EEC syndrome and cheilo-gnatho-palato-schisis had several congenital ear malformations. Syndromes connected with eye deformities were also observed. A boy with oral-cranial-digital syndrome also had several congenital malformations of the head, including hypertelorism.

TABLE 1

Child psychiatric hospital patients: 20 children according to diagnosis, gestational age and birthweight

Diagnosis			Gestational age*		Birthweight**		
Main group	Subgroup	Total	Term	Postterm	S.G.A.	A.G.A.	L.G.A.
Prenatal	Genetic	8	8		2	6	
	Acquired	1		1			1
	Unspecified	6	6		3	3	
Perinatal	Asphyxia	1	1		1		
Postnatal		3	3			3	
Psychosis		1	1			1	
Total		20	19	1	6	13	1

* Gestational age:
  Term: ≥ 37 < 42 weeks
  Postterm: ≥ 42 weeks
** Birthweight (B.W.):
  Normal: 2500–4500 g
  S.G.A.: small for gestational age, B.W. < 2500 after 37 or more weeks
  A.G.A.: average gestational age, B.W. and term within normal limits (not exceeding 2 S.D.)
  L.G.A.: large for gestational age, B.W. 4500 g or more

8. Tests to establish the level of development and functioning, including psychological (performance and verbal) tests, and special ones for language and motor development, and for social adaptation.

Although many of these examinations – somatic and psychological – were performed on all, or nearly all, cases, there is no routine screening programme. For each case, clinical examination provided the indications for further investigations. A final diagnosis or determination of the aetiology of the mental retardation was made according to the system used by Swedish investigators.[1,6-8]

Table 1 shows the results of the diagnostic process, and Table 2 gives a survey of the aetiological study.

TABLE 2

Child psychiatric hospital patients with mental retardation: the etiology according to chronology

Sex		AETIOLOGY		
		1. Prenatal		Total 15
		1.1 Genetic		
1M		Oral-cranial-digital	syndrome	
1M		Acrocephalosyndactyly	syndrome[2]	
	1F	Seckel's	syndrome[15]	
1M		Louis bar's	syndrome[13]	
1M		Marfan's	syndrome[14]	
1M		EEC	syndrome[17]	
1M		Pseudohermafroditism	syndrome[19]	
	1F	Rieger's	syndrome[16]	
		1.2 Prenatal (Less known inheritance)		
3M		Rubinstein-Taybi	syndrome[18]	
2M	1F	Unspecified		
		1.3 Prenatal aetiology (acquired)		
	1F	1.3.1. Toxoplasmosis		
		2. Perinatal		1
1M		2.1 Asphyxia		
		3. Postnatal		3
1M	1F	3.1 Cranial trauma		
1M		3.2 Lead poisoning		
		4. Chronology unknown		1
1M		4.1 Infantile psychosis		
		Total		20

DISCUSSION

Penrose (personal communication, 1959) stressed that aetiological factors responsible for severe mental retardation may also cause mild mental retardation. New diagnostic techniques (e.g. biochemical and chromosome analyses) have helped to elucidate such considerations. For instance, autosomal chromosome disorders are usually associated with severe mental retardation, and numerical disorders of the sex chromosomes frequently have relatively mild effects; in males with the fragile X syndrome, moderate retardation is common, but some are not retarded or only mildly so.

It is more difficult to establish these relationships in some mental retardation syndromes of unknown prenatal origin. Since no definite laboratory diagnostic tests are available for these syndromes, minimally affected cases with mild mental retardation are difficult to diagnose; facial traits and other signs may change with age (Fig. 1). Often the diagnosis of such a case is limited to 'mild

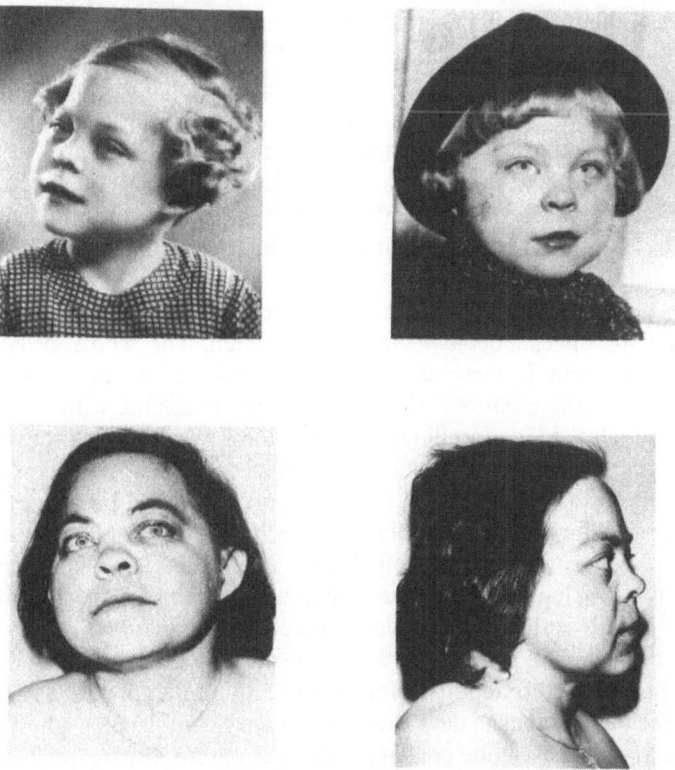

Fig. 1. Patient with Pseudohypotyreoidisme: 3 years, 6 years and 36 years old (this patient is not from the present analysis).

mental retardation of unknown aetiology' although a specialist experienced in dysmorphology may recognize the connection between the mild mental retardation and, for instance, the peculiar facies or congenital malformations.

It is very important to examine all these cases not only by 'simple screening', but by careful clinical appraisal supplemented by relevant laboratory and other tests. In our cases, if a chromosomal examination was deemed normal more than 5 years ago, a repeat study using banding and other current techniques is done where indicated. Initially, every patient is evaluated for manifestations of a specific syndrome by studying such features as height, body proportions, head shape, facial appearance, hair distribution (scalp and body hair), malformations of the extremities, and of the cardiovascular and other organ systems.

Many such cases remain classified as being of 'unknown aetiology' when no diagnostic evaluation is made by a medical specialist skilled in the diagnosis of mental retardation and malformation syndromes. Even if these syndromes are individually infrequent, they occur with some regularity among the mentally retarded: for instance, Cornelia de Lange syndrome, Rubinstein-Taybi syndrome, Prader-Willi syndrome, idiopathic infantile hypercalcaemia syndrome. Other examples are even more rare, including syndromes associated with ataxia, with ectodermal abnormalities (like the EEC syndrome), and with orofacial and/or digital malformations. A great help to the diagnostic team is an experienced ophthalmologist with special equipment to diagnose, for example, retinopathies.[20] A dentist interested in genetic syndromes is another very useful member of the team.[4] During psychological and child psychiatric examination it is important to observe for special myoclonic or psychomotor fits, such as those shown by a patient in the present series which were diagnosed as convulsions of the eyelids (Axenfeld).[16]

The typical case in the group reported here has, for many years, been treated in one of the special departments of a hospital. The whole situation is very difficult, especially in view of the many syndromes and diseases without any specific and pathognomonic sign. This creates problems with regard to treatment, and also in counselling respects. However, if a tentative syndrome diagnosis can be made, this will, in some cases, solve some problems. It may then be possible to tell the family that there is no well-known pattern of inheritance to enable specific genetic counselling. An example of a problem is a case of EEC syndrome where there was another case in the family. A comparison of the symptoms revealed that they were different, with some appearing in one case, and some in the other; nevertheless, there is a connection.

Also, for the child psychiatric team, a syndrome diagnosis may be useful if it can explain the combination of, for instance, mild mental retardation, difficulty with speech and language development, epilepsy, and peculiar physical development. In a recent case, mild epilepsy with psychomotor fits started before puberty and coincided with the birth of a brother, and a very severe jealousy reaction in the affected child. When the mother was told about the symptom pattern of that child's disease, which included psychomotor epilepsy, she asked, "Do you think it is not the jealousy reaction which caused the epilepsy?" When told about the natural history of the disease and cases with prepubertal onset, she said: "May I take it that even if he had not got a brother at that time, he would still have developed his epilepsy?" This could be confirmed, to the mother's relief.

With a child psychiatric approach heavily stressing early intervention and early stimulation and a close connection between the attitude of the parents and development of the child, it is often a great relief for the parents to know that the child's psychomotor retardation is a trait of a well-known disease, and not simply the result of either their particular family conditions or the way they have trained and educated that child. This is especially important in children with progressive mental/motor retardation. A good example is a child seen in our psychiatric department for psychosis and epilepsy of unknown aetiology, who, at the age of 10 years, started to develop the typical facial eruptions of tuberous sclerosis; this diagnosis facilitated family, including genetic, counselling. Even though skin depigmentation in tuberous sclerosis may be manifest at an earlier age, physicians are not always aware of this sign, and hence may not look for it. In mild cases of pseudohypoparathyroidism[10], the peculiar facial appearance is often reported in the medical records, sometimes with other associated signs. Pronounced asthenia causes severe problems for these patients in puberty and adulthood; they are unable to fully participate in either school or vocational training.

In 10 of the 20 cases in the present series, more than one aetiological factor was found. Establishing their precise chronology was not always possible. For example, in a case with both parents abusing drugs and the mother on therapy during pregnancy, the child had a complicated birth, and at the age of 9 weeks suffered a cranial fracture with severe brain damage. Records of this particular case suggested that postnatal brain damage was the most important factor, without which the child would probably have functioned at a low normal, or borderline, level of intelligence.

In those with genetic or acquired prenatal syndromes, complicated pregnancy and birth, small gestational age and/or neonatal problems were often present. In 1 severe case of lead poisoning, the child had pica and other signs of abnormal development prior to the poisoning. In 2 cases of severe psychosis, 1 had some somatic abnormalities suggestive of Marfan syndrome. In the 3 cases of Rubinstein-Taybi syndrome, the indications for hospital admission included abnormal, weak contact, aggressiveness and hyperkinetic symptoms.

Although only 1 case (with late onset infantile psychosis) was clearly progressive, the prognosis for all 20 is poor. A few of the cases were referred to special day institutions or schools for the handicapped, but the majority were integrated into the normal school system.

164

CONCLUSIONS

The cases reported here passing through a child psychiatric department constitute a rather special group of mentally retarded children. Their age range is wide, and diagnostic undertakings are often rather delayed because psychotic and/or behavioural symptoms dominate. As mental retardation is caused by so many different factors, it can be that some special groups of retarded children show symptoms which will direct them through child psychiatric departments when the referral system is based on symptomatology, which is the case in many Western countries. The symptoms in the group discussed included: speech difficulties, behavioural abnormalities (aggressive behaviour was common), special psychotic features (with difficulties in the contact with peers and adults and eventually also with the family) and other symptoms often seen among the mentally retarded (e.g., stereotype perseveration). It is surprising how often syndromes initially described in severely mentally retarded individuals are seen in milder forms among child psychiatric cases.

BIBLIOGRAPHY

1 Blomquist, H. K. son.: 1982, *Mental retardation in children. An epidemiological and etiological study of mentally retarded children born 1959–1970 in a northern Swedish county*, Umeo/a University Medical Dissertations, New Series, 76.

2 Chotzen, F.: 1932, *Eine eigenartige familiäre Entwicklungsstörung (Akrozephalosyndaktylie, Dysostosis craniofacialis und Hypertelorismus)*, Mschr. Kinderheilk., *55*, 97.

3 Dupont, A.: 1975, *Mentally retarded in Denmark. An epidemiological study of 21,000 registered cases. Some results of a census, May 1974*, Dan. Med. Bull., *22*, 243.

4 Dupont, B., Dupont, A., Bliddal, J., Holst, E., Melchior, J. C. and Ottesen, O. E.: 1970, *Idiopathic hypercalcaemia of infancy. The elfin face syndrome*, Dan. Med. Bull., *17*, 33.

5 Egekvist, H. ed.: 1982, *Fra gamle bakkehus til grønne skoler. Historiske artikler og kildeskrifter om åndssvageskolens 125-årige historie og dens forhistorie*. Åndssvageforsorgens Lærerforening, forlaget S. Å.-Materialer, Cophenhagen.

6 Gustavson, K.-H., Hagberg, B., Hagberg, G. and Sars, K.: 1977, *Severe mental retardation in a Swedish county II. Etiologic and pathogenetic aspects on children born 1959–1970*, Neuropädiatrie, *8*, 293.

7 Gustavson, K.-H., Holmgren, G., Jonsell, R. and Blomquist, H. K.son.: 1977, *Severe mental retardation in children in a northern Swedish county*. J. Ment. Defic. Res., *21*, 161.

8 Hagberg, B., Hagberg G., Lewerth, A. and Lindberg, U.: 1981, *Mild mental retardation in Swedish school children II. Etiologic and pathogenetic aspects*, Acta Paediat. Scand., *70*, 445.

9 Heron, A. and Myers, M.: 1983, *Intellectual Impairment. The Battle against Handicap*, Academic Press, London.

10 Illum, F., Dupont, E., Dupont, A., Jensen, S. B., Frederiksen, P. K., Konstantin-Hansen, K. K. and Lassen, L. B.: 1980, *Pseudohypoparatyreodisme, Nyere diagnostiske aspekter*, Ugeskr. Læg., *142*, 2606.

11 Jansen, J.: 1982, *Head Circumference in Danish Children. Allometric Growth*, Lægeforeningens Forlag, Copenhagen.

12 Jørgensen, O. S., Nielsen, K. B., Isager, T. and Mouridsen, S. E.: 1984, *Fragile X-chromosome among child psychiatric patients with disturbances of language and social relationships. A pilot study*, Acta Psychiat. Scand., *70*, 510.

13 Louis-Bar, D.: 1941, *Sur un syndrome progressif comprenant des télangiectasies capillaires cutanées*

et conjonctivales symétriques, à disposition naevoide et des troubles cérébelleux, Confin. neurol. (Basel), *4*, 32.

14 McKusick, V. A.: 1972, *Heritable Disorders of Connective Tissue,* 4th ed., Mosby Co., St. Louis, 61–223.

15 McKusick, V. A., Mahloudh, M., Abbott, M. H., Lindenberg, R. and Kepas, D.: 1967, *Seckel's bird-headed dwarfism,* New Engl. J. Med., *277*, 279.

16 Price, D. C. M. and Trounce, D. Q.: 1973, *Cyclic oculomotor paralysis,* Arch. Dis. Childh., *48*, 881.

17 Rüdiger, R. A., Haase, E. and Passarge, E.: 1970, *Association of ectrodactyly, ectodermal dysplasia and cleft lip-palate. The EEC syndrome,* Amer. J. Dis. Child., *120*, 160.

18 Rubinstein, J. H. and Taybi, H.: 1963, *Broad thumbs and toes and facial abnormalities. A possible mental retardation syndrome.*Amer. J. Dis. Child., *105*, 588.

19 Saez, J. M. and Morera, A. M.:1972, *Further in vivo studies in male pseudohermaphroditism with gynecomastia due to a 17-ketosteroid reductase defect,* J. Clin. Endocr., *34*, 598.

20 Warburg, M.: 1972, *Diagnosis of Metabolic Eye Diseases,* Munksgaard, Copenhagen.

A. CZEIZEL

THE PREVENTION OF MENTAL RETARDATION WITH THE HELP OF AN OPTIMAL FAMILY PLANNING PROGRAMME

The school-age prevalence (3.3%) and the distribution of the main causes of mental retardation for Hungary were known by the 1970s (Table 1) from genetic and terato-epidemiological studies.[2,15] Recently, the occurrence of recorded mentally retarded children has decreased to approximately 3%. This improvement was mainly achieved by preventive measures taken for some specific types of mental retardation, e.g. Down syndrome and phenylketonuria. Further progress can be expected through other specific preventive approaches, e.g. neonatal screening for hypothyroidism, and prenatal diagnosis of some X-linked types of mental retardation.[3,7] However, it is not clear whether hitherto available methods of qualitative family planning will enable a reduction of the so-called 'random risk', i.e. for the prevention of mental retardation in general. The random risk in Hungary during the past 14 years has been quite sizeable for unsuccessful pregnancy outcomes[1] and untoward conditions of offspring[4] (Table 2). Furthermore, 84% of congenital anomalies occur in the offspring of healthy couples with a negative family history, and this also applies to the different groups of pathological mental retardation. Recently, a number of young couples – the 'family planners' – have done their best to have healthy babies.

These considerations prompted us to establish an Optimal Family Planning Programme (OFPP) for the reduction of random risk and some specific anomalies. The OFPP was launched on 1 February, 1984.

PARTICIPANTS AND PROCEDURES

Couples who satisfy all 5 of the following criteria are eligible for participation in the OFPP:
– age between 18 and 35 in female family planners,
– no previous unsuccessful pregnancies (induced abortion is not an excluding factor),
– no delayed conception or infertility (more than 12 months of sexual activity without contraception),
– no pregnancy as yet,
– voluntary participation and promise of compliance with OFPP regulations.
  The flow chart of participants is as follows:
First meeting: (N = 2,004) Checking-up on reproductive health. Beginning of the 3 month preparation for conception.
Second meeting: (N = 1,114) Ending of the 3-month preparation for conception. Start in the achievement of conception or, if there is no conception, further second-type meetings II/b, II/c, II/d), as necessary.

167

E.K. Hicks and J.M. Berg (eds.), The Genetics of Mental Retardation, 167–175
© 1988 *by Kluwer Academic Publishers.*

TABLE 1

Distribution of categories and main causes of mental retardation in Hungary

Categories and causes	Mental retardation: Birth prevalence per 10^4 births	Percent		Severe/ < 50 IQ/ cases Percent	
Mendelian					
Inborn errors of metabolism (e.g. phenylketonuria, mucopolysaccharidosis types I, II, 111)	9.9	3.3		5.4	
Congenital anomalies (e.g. severe microcephaly, tuberous sclerosis)	11.1	3.7		8.4	
Fragile X syndrome	12.9	4.3	14.1	8.4	27.0
X-linked "non-specific" mental retardation without fragile X	6.0	2.0		3.4	
Congenital hypothyroidism	2.4	0.8		1.4	
Chromosomal					
Down syndrome	11.1	3.7		11.5	
Other autosomal aberrations	4.3	1.3	6.3	4.7	17.6
Sex chromosome aberrations (e.g. XYY, XXY)	4.3	1.3		1.4	
Prenatal					
Microbial (e.g. cytomegalovirus, toxoplasma, rubella)	5.1	1.7		4.7	
Fetal alcohol syndrome	25.6	7.7	9.6	2.0	6.7
Others (e.g. anticonvulsant drugs)	0.6	0.2		–	
Perinatal					
Birth trauma	3.3	1.1		1.0	
Hypoxia	59.4	19.8	21.8	32.1	34.5
Hyperbilirubinaemia	2.7	0.9		1.4	
Postnatal					
Meningoencephalitis	18.0	6.0	7.3	9.8	11.8
Others (e.g. trauma, tumour)	3.9	1.3		2.0	
Subtotal	180.6	59.1		97.6	
Familial*	122.4	40.8		2.4	
Grandtotal	300.0	100.0		100.0	

* See Myerson, A., 1930, "Research in feeble-mindedness with special relationship to inheritance", Bull. Mass. Dept. Dis., *14*, 2.

169

TABLE 2

Pregnancy outcomes and untoward conditions of offspring in the Hungarian population and in participants of the Optimal Family Planning Programme

Variables studied	Hungarian population figures /%/		Optimal Family Planning Programme			
			Expected figures		Observed figures	
	1870–1979	1980–1983	%	/Decrease/	% +	/Number/
Pregnancy outcomes						
Ectopic pregnancies	0.3	0.5	0.4	/20%/	0.0	/0/
Miscarriages, recognised	13.1	11.5	9.2	/20%/	6.3 ± 12.2	/37/
Stillbirths	0.9	0.8	0.7	/12.5%/	0.0	/0/
Livebirths	85.7	87.2	89.7	–	87.8	/266/
Total	100.0	100.0	100.0	–	100.0	/303/
Untoward conditions of offspring						
Birth weight under 2500 g	11.0	10.1	5.0	/50%/	2.7	/7/
Congenital anomalies /CAs/	6.0	6.0	5.4	/10%/	1.1	/3/+ +
including major CAs	2.3	2.3	2.0	/13%/	0.8	/2/
Infant deaths	3.1	2.1	1.6	/25%/	1.5	/4/
Mental retardation	3.3	3.0	2.0	/33%/	–	–

+ 303 completed and 286 ongoing pregnancies (i.e. 589 total pregnancies)

+ + one case had large omphalocele, prenatally diagnosed and pregnancy terminated by elective abortion; second case had cleft lip and palate; third case had an intrauterine healed cleft lip. All CAs were in placebo group

Third meeting: (N = 589) Pregnancy confirmation. Protection of very early pregnancy.

Fourth meeting: (N = 423) Evaluation of the pregnancy history until the 12th week of gestation. Pregnant women are referred to the prenatal outpatient clinics. Prenatal care. Delivery or other events, with information conveyed to the OFPP about pregnancy outcome (N = 319).

Last meeting: (N = 84) Checking-up 'optimal' babies at the 8th month of life.

As of 31 December, 1985, the number of eligible participants was reduced to 2,004 from 2,840 (836 couples were excluded because they did not meet the criteria).

The OFPP has 3 guiding principles:

I. Checking-up on reproductive health.

This first principle is based on 7 points:

1. Genetic background (family history). If severe monogenic or polygenic diseases occur in first degree relatives of family planners, the couple is referred to our genetic counselling clinic. This occurred in 6.2% of the participants.

2. Health condition (case history of female family planners). Some maternal diseases (e.g. epilepsy, diabetes mellitus, hyperthyroidism) require special pre-conception preparation and treatment during pregnancy by consultant experts. Maternal diseases were found in nearly 20% of the females.

3. Psychosexual condition (exploration). If either too infrequent or too frequent sexual activity occurs, we suggest that these activities be modified in the preconception phase. Other problems (e.g. ejaculation precox) needed counselling and treatment of 2.2% of couples in our sexological counselling clinic.

4. Pregnancy fitness (special gynaecological examination). Its purpose is to detect congenital anomalies and inflammation of the genitalia, as well as hormonal dysfunction based on phenotypic traits. Some treatment was necessary in 17.9% of females.

5. Male procreative fitness (voluntary sperm analysis). Two-thirds of male family planners provided semen for this examination. Surprisingly, in 17% of them a 'seminal weakness' with sperm cells below 20 million per ml. was found. They were treated in our andrological clinic.

6. Occupational background (occupational history).

7. Exclusion of some other risk factors (voluntary blood examination). Antibodies against rubella virus and toxoplasma gondii are studied and a check for anaemia is done. Rubella seronegative females (8.1% were found) are vaccinated. Unfortunately, Hungary has no population-based rubella vaccination programme. The prevalence of anaemia was 7.7% in females studied.

II. A 3 month preparation for conception

This second principle involves an additional 4 points:

8. Protection of germ cells. Both females and males are asked to avoid smoking, drinking alcohol, unnecessary medication and occupational hazards. There is a

high compliance rate with these difficult life-style proposals among our female family planners.

9. Restoration of hormonal balance after contraceptive pill use. The most common contraceptive method is oral anticonception (45%). We suggest discontinuing the use of contraceptive pills during this 3 month preparation period. Condoms are provided.

10. Determination of optimal days of conception by basal body temperature. We ask that this be measured for these 3 months in order to detect hormonal dysfunction (4.5% of females had anovulation and/or corpus luteal insufficiency; they were treated). We can estimate the optimal days of conception (i.e. day of ovulation and one preceding it) in 90% of females.[6] After the 'start' the couples were asked to aim for these 2 days, following 2 or 3 days of sexual abstinence.

11. Preconception multivitamin supplementation. The Elevit Pronatal[R] (Roche) and placebo (involving minerals and trace elements) are given according to a double-blind protocol for at least a month before planned conception. The main purpose is to check the efficacy of periconception multivitamin supplementation in the prevention of first occurrence of neural tube defects.[12,13] A complementary goal is to evaluate other possible benefits and harmful effects of this treatment for mothers and for fetal development. Maternal side-effects are rare and mild.

After the second meeting (i.e. at the end of the 3 month preparation period), nearly 92% of the couples are encouraged to commence conception. Female family planners are asked to visit us as soon as possible after the first missed menstrual period.

III. Protection of very early pregnancy

This third principle of the OFPP is concerned with 4 further points:

12. Very early diagnosis of pregnancy. The basic aspect of the third principle is a very sensitive immunological pregnancy test based on the beta-component of hCG which can detect pregnancy on approximately the 10th–14th day of gestation.[8] At present, in general, pregnancies are diagnosed by obstetricians at about the 6th–8th week of gestation, i.e. after the critical period for major congenital anomalies. Two hours after the blood sampling at the third meeting, pregnancies of participants are confirmed or excluded. This blood sample is also adequate for the study of an antianaemic effect of multivitamins taken prior to conception. Pregnancy diagnosis is supplemented with an ultrasound examination in the 4th–6th week of gestation. The use of optimal days for conception shortens the period needed to achieve pregnancy (3 months vs. 3.9 months in the general population). The early recognition of pregnancy has some considerable advantages.

13. Avoidance of hazards to the fetus. Females become aware of their pregnancies at a very early stage, and this helps them to protect the fetus from potentially dangerous environmental factors. Since it is well-known that the beginning of the critical period for congenital anomalies is about the 14th day of gestation, the fetus is protected as early as possible. Pregnant women are

exempted from occupational hazards, if applicable, from the time of the first missed menstrual period. This occurred for 3.1% of female family planners.

14. Postconception multivitamin supplementation. This continues until the 4th meeting, i.e. until the 12th week of gestation, calculated from the last menstrual period.

15. The use of further protection methods. The importance of available prenatal screening (e.g. maternal serum alphaafetoprotein examination and ultrasonography) is stressed and some advice on how to protect the fetus in the further period of pregnancy is given, both orally and in printed form. A further voluntary blood test is done for the detection of seroconversion in toxoplasma antibodies and to evaluate any improvement of anaemia due to the multivitamin treatment.

After the fourth meeting, our participants are referred to the regional prenatal outpatient clinic. They get a certificate form from the OFPP which we get back at the end of the pregnancy, this serving as a report on pregnancy outcome. The certificate is filled out by the mother, but confirmed by the obstetrician. As a reward for obtaining it, we send the family a present – a baby carrier. Some cases are lost to follow-up, but one of our co-workers visits these couples in order to clarify their pregnancy outcomes.

Eight months later our couples are invited, with their infants, for a detailed check-up which involves the examination of:

1. somatic growth by anthropometric measurements;
2. paediatric status, including observation of minor developmental variants;
3. sense organ function;
4. behavioural development using the Budapest Developmental Test compiled by Szegal.[14] This test gives 5 indices: Fine-Motor and Gross-Motor Development indices and indices for Verbal Development, Social Competence and Social Skills. A general Developmental Quotient (DQ) can be calculated using the first 4 indices. The fifth index has a function similar to the Vineland Social Maturity Scale. Although certain unevenness of development should be viewed as physiological, a considerable difference in scores for different facets of development may be an early sign of potential pathology.
5. Mental development by the Popper-Szondy Functional Development Test which assesses 10 variables of motor, verbal and social development. Its general score is compared to data from the relevant general population.

Finally, a very detailed assessment of the parents is made at this early evaluation, on the basis of a parental attitudes questionnaire (PARI)[11], the Life Experiences Survey[10], parental IQ testing (Raven test)[9], the Brengelmann Questionnaire(a West German version of the Eysenck Questionnaire)[5], and an observational test of the attachment. At that time families get another present – 2 baby-sets.

'Control' infants are studied in blindfold fashion, using the same method. They are selected from the same inpatient obstetric clinics where the index babies were born and matched with the latter for sex, month of birth, and educational level of the parents.

The preliminary results of pregnancy outcomes in the OFPP are encouraging (Table 2). Particularly noteworthy are the lower than expected observed figures of untoward conditions of offspring. The mean birth weight of 'optimal' babies significantly exceeds that of the Hungarian general newborn population during the study period (Figure 1).

BIRTH WEIGHT DISTRIBUTION

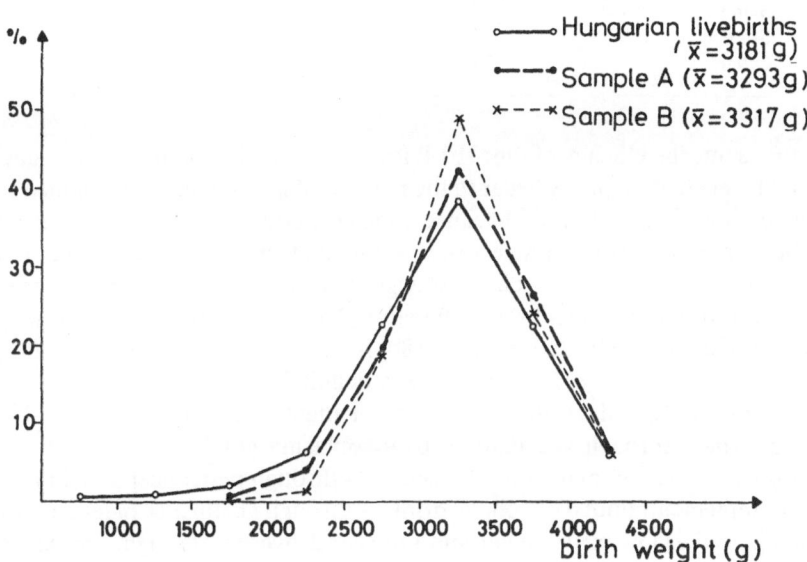

SAMPLE A: periconceptional use of Elevit Pronatal
SAMPLE B: placebo

DISCUSSION

Of course, the costs and benefits of the OFPP can only be adequately evaluated on a much broader scale. In addition to statistical problems, we are aware of difficulties in separating biological from social factors in the data. The social class distribution of our participants does not correspond to the distribution of the Hungarian pregnant population in general (Table 3). This is also apparent from the data on life-style. All types of unsuccessful pregnancy outcomes, including untoward conditions in offspring, are negatively correlated with parental socio-economic status. Since the socio-economic and age specific figures of the variables are known to some extent, we hope to be able to estimate the proportions of social and biological factors. All these factors have to be taken into account in the calculation of expected figures from the OFPP (Table 2).

TABLE 3

Social class distribution of female participants in the Optimal family Planning Programme and in the Hungarian pregnant population in the 1980s

Social class	Female participants (%)	Hungarian pregnant population (%)
Intelligentsia	51.0	10.2
Clerk	20.1	23.0
Skilled worker	20.2	36.1
Semiskilled worker	1.4	15.6
Student	3.5	0.2
Others	3.8	14.9

At present, the efficacy of the OFPP for the prevention of mental retardation cannot be evaluated, partly because the number of index children examined thus far is too low (84), and partly because a longer period of observation is needed to obtain reliable data. An arrest of development may be detected only much later, when the child enters school, though its cause may, in fact, be inborn, as in the case of minimal cerebral dysfunction. However, 2 points may be noted: no cases of definite retardation (DQ < 80) have to date been identified, and the average score for the group as a whole is definitely over 100.

Our hope is to reduce the occurrence of mental retardation caused by a) rubella virus – through vaccination; b) toxoplasma gondii – through prenatal serological screening; c) neural tube defects – through periconceptional multivitamin supplementation; d) alcoholic drinks – through abstinence of females from drinking; and e) low birth weight and preterm delivery – through, for example, the cessation of smoking.

It is an important point that couples with previous unsuccessful pregnancies are excluded from the OFPP. They are asked to visit genetic counselling clinics. Unfortunately, parents with familial mental retardation also do not take part in the OFPP. However, the experience of qualitative family planning methods may contribute to the improvement of general health concepts and thus have a beneficial effect on further offspring.

BIBLIOGRAPHY

1 Czeizel, A., Bognár, Z. and Rockenbauer, M.: 1984, *Some epidemiological data on spontaneous abortion in Hungary, 1971–1980,* J. Epidem. Com. Hlth., *38,* 143.

2 Czeizel, A., Lányi-Engelmayer, A., Klujber, L., Métneki, J. and Tusnády, G.: 1980, *Etiological study of mental retardation in Budapest, Hungary,* Amer. J. Ment. Defic., *85,* 120.

3 Czeizel, A. and Métneki-Bajomi, J.: 1978, *Present possibilities for the prevention of hereditary mental subnormality,* Int. J. Rehab. Res., *1,* 301.

4 Czeizel, A. and Sankaranarayanan, K.: 1984, *The load of genetic and partially genetic disorders in man I. Congenital anomalies estimates of detriment in terms of years of life lost and years of impaired life,* Mut. Res., *128,* 73.

5 Eysenck, H. J.: 1965, *The questionnaire measure of neuroticism, extraversion and introversion*, Rivista Psichol., *54*, 113.

6 Hertig, A. T., Rock, J. and Menken, M.: 1959,*Thirty-four fertilised human ova, good, bad and indifferent, recovered from 210 women of known fertility. A study of biological wastage in early human pregnancy*, Pediatrics, *23*, 202.

7 Lányi-Engelmayer, A., Katona, F. and Czeizel, A.: 1983, *Current issues in mental retardation in Hungary*, Res. Ment. Retard., *4*, 123.

8 Pácsa, A. S., *Measurement of human chorionic gonadotrophin – tested by Audin-Biotin-enzyme-linked-immunosorbent assay: detection of early trophoblastic activity in relation to ectopic pregnancies*, Gynec. Obstet. Invest., in press.

9 Raven, J. C.: 1956, *Guide to Using Progressive Matrices*, London University Press, London.

10 Sarason, J.G.: 1978, *Assessing the impact of life changes: development of the Life Experiences Survey*, J. Consult. Clin. Psychol., *46*, 932.

11 Schaeffer, E. S. and Bell, R. Q.: 1958, *Development of a parental attitude research instrument*, Child Develop., *29*, 339.

12 Smithells, R. W., Nevin, N. C., Seller, M. J., Sheppard, S., Harris, R., Read, A. P., Fielding, D. W., Walker, S., Schorah, C. J. and Wild, J.: 1983, *Further experience of vitamin supplementation for prevention of neural tube defect recurrences*, Lancet, *1*, 1027.

13 Smithells, R. W., Sheppard, S., Schorah, C. J., Seller, M. J., Nevin, N. C., Harris, R., Read, A. P. and Fielding, D. W.: 1980, *Possible prevention of neural tube defects by periconceptional vitamin supplementation*, Lancet, *1*, 339.

14 Szegal, B.: 1980, *Diagnostics of psychomotor development*, Magyar Pszichológiai Szemle, *37*, 148.

15 Vitéz, M., Korányi, G., Gønczy, E., Rudas, T. and Czeizel, A.: 1984,*A semiquantitative score system for epidemiologic studies of fetal alcohol syndrome*, Amer. J. Epidem., *119*, 301.

M. A. M. DE WACHTER

RECENT DEVELOPMENTS IN GENETIC COUNSELLING: ETHICAL VIEWPOINTS

GENETIC COUNSELLING AND ITS GOALS

The essence of genetic counselling is "to inform couples at risk of having offspring with inherited disorders or congenital malformations about this risk and to discuss with them various aspects of the different alternatives for action, particularly with respect to family planning."[1] All counselling services will agree with this. How to present this information and exactly how to discuss the options and choices is, however, still open for debate. Additionally, the content of the genetic counselling itself is more complex than the above definition suggests.

Fraser and Challis[6] described 4 areas or phases in the genetic counselling process: 1) estimation of recurrence probability, 2) informative counselling, 3) supportive counselling for the resolution of emotional conflicts, and 4) a follow-up for counselees and relatives. If the 'freedom' of the counsellees is not a prior assumption, the latter phase might suggest a possible tension between the interests of the individual counsellees and those of society. Shaw[15] reflects this tension in her listing of 18 objectives of genetic counselling related to the three parties involved, i.e. the affected individual, the parents, and society. This list is neither exhaustive nor generally agreed upon by geneticists.[14] Even if non-directive counselling is generally accepted, it remains unclear whether couples are both encouraged to make, and supported in making their own decisions.

In the Netherlands, as elsewhere, geneticists are aware of the complexity of goals and objectives of genetic counselling. It aims "at providing all information necessary for the counselee in order to make his/her own reproductive decision in accordance with that person's own view of life."[8] Therefore, counsellors abstain from giving their own private opinions. Yet, in order to assist in this decision process, the counsellors are obliged to provide extensive medical genetic information. Also discussed will be the prognosis of the relevant disease and various reproductive options such as prenatal diagnosis, artificial insemination by donor, anticonception, sterilisation, and adoption. This information process may be consciously and/or unknowingly biased by the counsellor's impressions of the attitudes of the counselees.

THE FABRIC OF ETHICS IN GENETIC COUNSELLING

Both the definition of genetic counselling and its objectives are fraught with ethical issues. The professional community of medical geneticists has been the

E.K. Hicks and J.M. Berg (eds.), The Genetics of Mental Retardation, 177–184
© *1988 by Kluwer Academic Publishers.*

first to recognise these issues, and to begin the search for acceptable solutions. Confidentiality of information, for example, remains a guiding principle. However, from the early days of counselling, conflicts have been experienced between this principle of privacy and the rights of relatives to know the facts related to their reproductive risks. As a solution, Fraser and Challis[6] proposed that the counselling process should be "not only person-oriented, but people-oriented."

The rapid development of clinical genetics and the complexity of new genetic knowledge are other reasons for ethical considerations. Presymptomatic diagnosis of disorders of late onset (e.g. Huntington disease) raises the ethical dilemma of the application and disclosure of the ensuing information. If this is not given now, it must be decided through what procedure one will secure a patient's right to know.[1] Again, there is an important privacy aspect here: what if information about persons with such diseases should leak into public files (e.g. those of the labour market and job opportunity organisations)? Mobilising current moral wisdom of geneticists in establishing guidelines will help "to meet more serious demands" of the future.[4]

Another point concerns the discrepancy between principle and practice. Currently, most centres for prenatal diagnosis make it clear to couples that the final decision about termination of an affected pregnancy lies with them when the test results are known. Sometimes, an apparent discrepancy is experienced when prenatal diagnosis is not given to parents who in principle refuse the abortion option. However, genetic centres also have an obligation to reserve scarce facilities for those couples who, at the outset of the procedure, state their wish to prevent the birth of an affected offspring. Careful counselling may prevent parents from lying about their objectives and may forestall unwanted and undesirable conflicts during mid-gestation.

Finally, ethical guidelines on genetic counselling are needed in pluralistic societies with their variety of ethical opinions. There is no general consensus on selective abortion, and on the moral status of the embryo. Some counsellors believe they should stay outside of this moral debate and leave it to the parents themselves, whereas other think they should assist parents in the decision-making process in accordance with the parents' views.

THE MAIN AREAS FOR MORAL CONCERN

The main areas for moral concern in genetic counselling are: 1) the content and distribution of information, how much and to whom; 2) counselling versus freedom of choice, in particular the issue of directive vs. non-directive counselling; 3) the problem of selective abortion; and 4) the possible tension between options of individuals and those of society ('eugenics' and prevention).

Another area receiving some renewed interest concerns paternity. The illegitimate child may be unaware of the real risk, or, on the contrary, may suffer from unfounded anxiety. Some geneticists argue in favour of the right to know one's genetic origin. However, such a right may, in practice, be very difficult to realise.

Given the taxing and apparently indiscrete questioning involved in determining paternity, other avenues of approach to determine paternity are preferable, e.g. obtaining such information during family studies, without prior knowledge of the counsultands, particularly when the information is crucial to, for instance, a new prenatal diagnostic test. This is already the case in the application of DNA techniques for carrier testing and prenatal diagnosis for diseases such as Duchenne muscular dystrophy, haemophilia, and cystic fibrosis. Counsellors cannot ignore these implications. They should seriously reflect, prior to disclosing any facts on the principle that the findings must be unavoidably disclosed to the patient who, in turn, must decide about further disclosure.[5]

Sometimes the ethical issues are compounded by the technical, medical and managerial problems of clinical genetics. Diagnostic uncertainty may increase the difficulties of telling or not telling the patient. Genetic heterogeneity, for example, whereby 2 clinically similar entities may show different modes of inheritance, cause both diagnostic and counselling problems, which may be difficult to explain. Moreover, clinical geneticists generally do not have the opportunity to revise a diagnosis during extended contacts with families, since these contacts are usually only on a short-term basis.[5] Since the consequences of these short contacts may be lifelong, this is another reason for geneticists to familiarise themselves with the moral sensitivities of their counsellees. A moral issue which often arises concerns the question of telling or not telling a counsellee. The question of when a counsellor withholds information must be addressed as well as the reasons for withholding it. Before considering such specifics, the general frame within which such medical ethics may work is briefly outlined below.

A FRAME OF ETHICS

Any ethicist may identify him/herself with one of the many ethical systems, e.g. rule-utilitarianism, consequentialism, deontologism. Ethical judgements may also be based on a careful consideration of three dimensions of action (the specific intention, the act itself, and the consequences). Each of these dimensions may be judged according to the generally accepted principles of medical ethics of non-maleficence, beneficence, autonomy, and justice.[11] An adequate judgement can be reached by checking the current meaning of each principle, e.g. autonomy as a moral right of a person may involve more than purely legal rights of the individual; consequently, moral obligations will be different from legal ones.

Since genetic counselling is a process of communication, one must define the relationship between counsellor and counsellee. Should the counsellor relate to the counsellee in a purely instrumental way (i.e. here are the facts, you decide!)?; or should the relationship be a conventional one [2,13]; or should it be a functional relationship (i.e. 'relative to your free procreative decision, here is the relevant information')?; or, finally, should the relationship be personal (i.e. 'since I commit myself to your well-being, this is what I consider in your best interest')? The counsellor must choose a role, and perhaps make the counsellee aware of the chosen model qualifying the relationship. More research in this area may clarify a number of problems in counselling processes.

An effort is made in the following to apply the above-mentioned dimensions to the ethical issue of withholding information in the counselling situation, especially when there is doubt about the clinical facts or whether the knowledge will be beneficial to the counsellee. It is this author's opinion that a functional relationship offers the best chances for adequate counselling. The implications of this preferred model will be developed in terms of telling the counsellee what one knows and how well one knows it.

The easy situations are the ones where clear agreements have been reached throughout the counselling sessions, e.g. with regard to unexpected findings. Here, it is easy to recommend absolute honesty with parents, the sharing of uncertainty, and to adhere to the principle that parental autonomy prevails. Concerning uncertain diagnosis, however, there may arise difficult dilemmas where new decisions will have to be made with regard to communication of information. These are situations where the wisdom of professionals is being tested. It is not unwise, in these situations, for a counsellor to rely on the wisdom of the group, i.e. the profession which is by now well aware of the fact that an uncertain diagnosis is a heavy burden on parents. Also, if uncertainty increases people's anxiety, it is unlikely that their decision-making would be unimpaired. Should one then rightly withhold all possibly harmful, or even burdensome, information from patients? Would it be appropriate for a genetic counsellor to say, for example, that in the case of fetuses aborted because of a prospect of having an X-linked disorder which cannot be specifically identified *in utero*, all parents should always be told that the fetus was severely affected, no matter what the actual findings had been? What is the ethical rationale for such decisions? And, how will the above suggested frame of ethics apply?

THE ETHICAL DILEMMA OF COUNSELLING WITH UNCERTAIN INFORMATION ABOUT RISK

For Down syndrome and other chromosome disorders the question arises as to whether the risk estimation should be based on birth survey data, or on data obtained on the basis of amniocentesis.[16] In other examples recurrence risks may be more difficult to ascertain, and become the subject of personal interpretation by the counsellor. Interpretation of facts and conclusions about risks may become subjective. Obviously, the scientific analysis of data is a primary task of the clinical geneticist/counsellor. However, the counsellor may be knowingly or unknowingly biased. His or her preference and willingness to accept risks may differ from those of the counsellee.[3] If, after all this, the counsellee asks the counsellor: "Doctor, what would you do?", then the counsellor may be tempted to opt for directive counselling. Directive counselling is, however, rejected by virtually all counsellors. And yet, why is it wrong, when people ask for it?

In this area there are substantial differences between American and European biomedical ethics. In Europe, the more traditional pattern of a relationship prevails, whereas this author perceives counselling in the United States as being more observant of contractual notions of right and autonomy, and deeply influenced by the feared malpractice suit. This will lead to substantial differences in counselling when only sparse information is available. If this cultural difference in fact exists, then international professional groups should be stimulated to develop a more comparative view.

Is it true, as often postulated in the literature, that informing the patient solves many problems? Does this apply to certain and to uncertain information? Information may contain definite facts, like the small risk of a spontaneous abortion after amniocentesis. On the other hand, the prenatal diagnosis of some sex chromosomal abnormalities will lead to problems in decision-making, i.e. the features of these disorders are highly variable and mental retardation is present in a limited number of cases. Thus, no precise prediction of the outlook for the fetus or child is possible. The parents are left with the ethical problem of how to make reproductive decisions or decisions about terminating a pregnancy based upon this type of information. Is there a moral limit to the possible use of 'uncertain' information? When does information become counterproductive in the exercise of rational human freedom? The issue is ethical as well as medical and technical. Ethical theories have not fully addressed the problem of 'rational and free decisions' when only limited information can be offered.

In such situations conflicts may arise between informative and supportive counselling roles, and, for individual counsellors, the boundary between advisor and decision-maker may become quite narrow. This seems far removed from the ideal of non-directive counselling where, for example, "the decision whether to terminate a pregnancy must be taken by the parents on the basis of adequate information from the counsellor."[16] However, it is not always clear what this 'adequate information' entails. Sometimes, genetic counselling seems to be the art of helping people to decide with minimal information.

THE DILEMMA OF COUNSELLING WITH CERTAIN YET POSSIBLY HARMFUL INFORMATION

Within a 'functional relationship' (i.e. where the services of the counsellor function as an information base in the counsellee's decision-making process), a counsellor ought to offer firm and straightforward professional advice, mitigated by a genuine receptiveness to the counsellee's wishes and his or her real needs. The concern for the counsellee's best interest justifies, as previously indicated, the withholding of uncertain information on the ground that such information either may be harmful or of no use. But what should one think of certain information being withheld from counsellees? Even more, what should be said about withholding certain information from counsellees for their best interest?

Traditional ethics distinguish between *falsiloquium* and *mendacium*. The former is a way of speaking which does not correspond to the actual facts. This form of not telling the truth does not necessarily have a moral connotation. The latter, on the other hand, does have moral significance. It is an untruth not because it is speaking against true facts but because there is an obligation to speak the truth. With this distinction in mind, one may ask whether there are good reasons that would justify a counsellor's information not being truthful. In other words, when does the obligation to speak the truth cease to exist? One reason for suspending the obligation would be prior agreement with the counsellee about withholding certain kinds of information. All other reasons for withholding information are similar to the reasons applied to imparting uncertain information, viz. they apply the principles of non-maleficence and beneficence. Indeed, some clinicians and counsellors say that they perceive a counsellee's inability to carry the moral burden of full genetic information. Others argue that patients are entitled to receive psychosocial encouragement, both to protect them from suffering and to offer them a climate of hope. It is also being said that patients deserve moral support for a choice they have already made. Finally, counsellors have a duty to contribute to the optimal human climate in order that their patients can go on living well and be able to consider new procreative choices. All of these choices for life justify for many counsellors their occasional withholding of information for the benefit of patients.

Convincing though these arguments may be for many, it is understandable that others remain doubtful about their ethical justification. Should one acknowledge such withholding of information as right, it would then be wise to consider the following: 1. the context should be one of best interest of the particular patient; 2. it may always be best to consult colleagues before counselling a patient in an untruthful way; and 3. if a given counsellor often thinks it right to withhold information, that counsellor may not be truthful to himself.

TRADITION OR CONSENSUS?

Basically, there are two possible ways out of ethical dilemmas, i.e., to deal with them with traditional answers, or by building a consensus. Should the genetic counselling session not be able to offer innovative medical solutions, then traditional medical ethics would offer a useful basis for patient-physician relationships. Classically, physicians and patients have always made decisions between alternatives in order to "find the solution that is the lesser evil."[1] In genetic counselling, it leads to the responsibility of patients with severe genetic disorders, like Huntington's chorea, to prevent their spread. How could one ethically defend not accepting information? As a consequence, there is every right to offer (impose) unsolicited counselling. Past experience seems to justify this approach. Examples of actively pursuing contacts can be found in the field of infectious diseases (it was and is quite common to contact people who might possibly spread or be infected with disease). Tuberculosis, venereal diseases, and AIDS all are examples where there is the wish to prevent further spreading of the disease. Traditional medical ethics are undoubtedly valuable in solving these current problems. In opting for the lesser evil, the principles of double effect, probabilism, conflict and compromise have been applied. Through all of these efforts, ethicists have attempted to "limit the area of moral ambiguity in hard cases by specifying the objective criteria for determining to what extent we can cause evil in the pursuit of good."[10] Using the well-known examples of ectopic pregnancy and venereal disease, ethicists built a system of weighing values against the loss of values. The disadvantage of individual burdens were weighed against the gain of collective good health. The principles derived from such cases have been advocated as instruments to solve today's ethical problems in medicine.

The traditional approach has its limitations, however. Society has changed its expectations of medical technology. Patients, behaving as well-informed consumers, are no longer satisfied with the second best or the lesser of two evils. They now want nothing but the best. For their offspring, many people expect the prevention of all evil (negative eugenics) and, for the future, ever increasing quality (positive eugenics). Today, the increasing demand for prenatal diagnosis from groups not at increased risk reveals the fears of having an impaired child and the desire to exercise control over the quality of one's offspring.[12]

Tradition is, by definition, conservative. One may ask how a traditional medical ethics would keep up with rapid, complex and radical biomedical innovations. The principle of double effect was applied unsuccessfully in the abortion debates of the 1960s. Many ethicists have abandoned this principle for the ethical assessment of clinical situations. The gap between biomedical development and bioethics should, therefore, be a major concern of ethicists. In searching for an alternative, some ethicists have turned to consensus as a radically new way out of ethical dilemmas. For example, consensus was proposed by ethicists in the Netherlands on the issues of embryo research and of euthanasia.[7]

When no general agreement can be reached in a pluralistic society, all efforts to reach minimal agreements become meaningful. Such consensus regarding euthanasia would reduce tensions in confrontations with views based upon strong religious convictions. Eventually, consensus would enhance a rational policy in such delicate matters.

The danger of replacing ethics by policy making has been mentioned by many critics of some North American procedural ethics.[9] Regulation must never become a substitute for morality.

Another criticism of consensus building is that it may cause new tragedies. Minimal moral agreements tend to lower the moral level of the society as well as of individuals. This inherent paradox of consensus building constitutes a dilemma for those ethicists who believe they must choose either tradition or consensus.

Applied to clinical genetics, it is unlikely that a minimal moral consensus will solve fundamental disagreements on selective abortion or embryo research. It may, however, lead to solutions for practical problems in genetic counselling and thereby become more generally acceptable. For example, the right of access to prenatal diagnosis will be readily acknowledged and become a part of a practical consensus. The same would apply to some aspects of informing counsellees, their relatives, and others with vested interests about, for example, future marital relations. Similarly, it will be generally accepted that appropriate choices can be made only if adequate information has been provided. These examples illustrate the value of a moral consensus to delineate the moral good in clinical genetics. Since ethics has always been actively involved in the practice of clinical genetics, the chance for an integrated relationship between science and ethics is increased.

Fletcher[4] has pointed out that "Facing such issues with some help of guidelines is better than facing them simply with the strength of convictions." This obviously applies to clinical geneticists who will be facing increasingly complex issues. However, because it says very little about the inherent ethical quality of the guidelines, a basic task for ethicists working with consensus building must be to critically assess the ethics of consensus.

BIBLIOGRAPHY

1 Berg, K.: 1983, *Ethical problems arising from research progress in medical genetics*, In, K. Berg and K. E. Tranoy, eds., Research Ethics, Alan R. Liss, New York, 264–270.

2 Defares, P. B.: 1963, *Grondvormen van Menselijke Relaties*, van Gorcum, Assen.

3 Evers-Kiebooms G., Cassiman, J. -J., van den Berghe, W, and d'Ydewalle, G.: 1987, *Genetic Risk, Risk Perception, and Decision Making*, Alan R. Liss, New York.

4 Fletcher, J. C.: 1983, *The evolution of the ethics of informed consent*, In, K. Berg and K. E. Tranoy, eds., Research Ethics, Alan R. Liss, New York, 201.

5 Fletcher, J. C.: 1985, *Ethical aspects of medical genetics*, Clin. Genet., 27, 199.

6 Fraser, F. C. and Challis, E. B.: 1978, *Genetic counseling*, In, R. E. Razel and H. F. Conn, eds., Family Practice, 2nd ed., W. B. Saunders Co., Philadelphia, 465–478.

7 Fretz, L.:1985, *Mag embryonaal menselijk leven instrumenteel worden gebruikt?* In, NRC Handelsblad, 23 Feb.

8 Haar, ter, B. G. A.: 1979, *Genetic counseling*, In, A. C. Drogendijk, ed., Zorgen voor Morgen, Nijkerk, Intro., 86.

9 MacIntyre, A.: 1980, *Regulation: A substitute for morality*, Hastings Center Rep., 10, 31.

10 May, W.: 1978, *Double effect*, In, W. T. Reich, ed., Encyclopedia of Bioethics, The Free Press, New York, 317.

11 National Commission for the Protection of Human Subjects in Biomedical and Behavioral Research: 1978, The Belmont Report, U. S. Govt. Printing Office, Washington, D.C.

12 Nippert, I.: 1984, *Die Angst ein mongoloides Kind zu bekommen – oder Risikoverhalten und der Weg zur genetischen Beratung*, Med. Mensch Gesellschaft, 9, 111.

13 Oldendorf, A.: 1953, *Psychologie van het Sociale Leven*, Bijleveld, Utrecht.

14 Roy, D. J. and de Wachter, M. A. M.: 1986, *The Life Technologies and Public Policy*, The Institute for Research on Public Policy, Montreal, 68.

15 Shaw, M. W.: 1977, *Review of published studies on genetic counselling*, In, H. A. Lubs and F. de la Cruz, eds., Genetic Counseling, Raven Press, New York, 36.

16 Stene, J. and Mikkelsen, M.: 1984, *Down syndrome and other chromosome disorders*, In, N. J. Wald, ed., Antenatal and Neonatal Screening, Oxford, 74.

J. ZAREMBA

ETHICAL QUESTIONS IN COUNSELLING FOR PATIENTS AFFECTED WITH TUBEROUS SCLEROSIS

Tuberous sclerosis (TS) is a moderately rare autosomal dominant genetic condition. The prevalence in Poland, based upon data obtained from mentally retarded institutionalised patients, has been estimated as 1 in 23,000.[8] When mild cases are included, as British and American data show, the prevalence at birth may be as high as 1 in 10,000.[4,5] Mental retardation, epilepsy, typical skin changes and frequent involvement of internal organs (e.g., kidneys, heart) occur in the fully manifesting syndrome. However, the clinical picture varies considerably depending upon the extent and sites of the lesions. Thus there are severe cases and about 50% or more that are mild and abortive without mental retardation[4] (the latter often accounting for a familial incidence of the disease).

The penetrance of the TS gene is complete and the genetic risk to offspring when one parent is affected is 50%. However, approximately 75% of TS cases are sporadic and attributed to fresh mutations. This conforms to the well-known genetic rule that the more reduced the genetic fitness, the larger the proportion of fresh mutations. Indeed, in TS, the genetic fitness (roughly meaning the ability to reproduce) is greatly reduced.

Those affected with a severe form of the disease often attract attention, but do not generally constitute a very difficult genetic counselling problem because they are easy to recognise and their genetic fitness is close to zero.

More important is the ability to detect persons with mild and abortive forms of the disease since they are a potential source of familial cases. For such detection, one needs experience and knowledge of the disease, and modern diagnostic facilities such as computed tomography, magnetic resonance imaging, and echocardiography. Clinical diagnostic skills, supported by these sophisticated techniques, will enable a diagnosis to be made in a great majority, probably over 95%, of cases. Unfortunately, so far prenatal diagnosis of TS is not available and no genetic marker of TS is known, although there is evidence for a possible genetic linkage of the TS gene with the ABO blood group genes.[2] This has recently been confirmed by Fryer et al. who, by proving very close linkage between TS and the ABO blood group, were able to assign the TS gene to the distal long arm of chromosome 9.[3]

As multiplicity of organ involvement and variability of manifestations are characteristic of TS, detection of mild cases requires a very thorough investigation. A too superficial examination of a person genetically at risk for TS, or clinically suspected to have the condition, is neglectful. When findings typical for

185

E.K. Hicks and J.M. Berg (eds.), The Genetics of Mental Retardation, 185–187
© *1988 by Kluwer Academic Publishers.*

the disease are detected, the patient should receive detailed counselling about the genetic risk, as well as information on available prophylactic measures including the choice of contraceptives, selective abortion, and sterilisation. The choices and decisions should be entirely up to the patient and her or his partner.

Controversial may be the cases in which an affected person is mentally retarded, able to conceive but not quite able to understand the consequences. Various measures have been proposed for such instances. Borberg[1] quoted Curtius (1935) and Seidl (1940) as proposing sterilisation of all subjects affected with severe, as well as mild and abortive, forms of TS. Borberg,[1] writing in 1951, also referred to clause 10 of the Danish Marriage Act which made issue of a marriage license to an affected person subject to permission of the Ministry of Justice, and such permission could be conditional on sterilisation of the said person. Borberg considered that this provision should be enforced in patients affected with severe forms of TS, as well as in those with an abortive form who suffered from epilepsy. Such eugenic measures would now generally be considered as a violation of human rights. However, it seems ethical to suggest sterilisation as one possible prophylactic measure in some cases of TS, at least until a method of prenatal diagnosis becomes available. In abortive and mildly pronounced forms it should only be performed on a voluntary basis. In mentally retarded persons with TS, who are capable of reproducing but unable to comprehend and to cope with the consequences of it, sterilisation should perhaps be enforced, but in every case this should be based on a court decision, since it involves interdiction. Nevertheless, permission to marry should never be conditional on sterilisation, particularly as, in many present day liberal societies, marriage is by no means an indispensable condition for couples living together.

It should also be remembered that sterilisation is usually irreversible. As a genetic probe for the TS gene, and hence prenatal diagnosis, may soon be available, perhaps preventive measures other than sterilisation should be chosen, particularly in mildly affected persons.

Induced abortion should be offered in view of the 50% genetic risk, and high risk for a severe form of the disease, in offspring of a mildly affected person. In cases of an affected male, artificial insemination is another alternative. This again should be presented as a voluntary option to interested couples.

Members of a family in which the TS gene is present should always be informed about the possibility of being affected and about the genetic risk involved. Genetic counselling, clinical examination and all necessary diagnostic tests should be made freely available to them.

The ways in which affected persons are informed is very important. Particularly those affected with mild forms of the disease, of which they have never before heard, often find it hard to understand and believe that a few nodules or skin patches, for instance, can be associated with a 50% risk of having an abnormal child. Therefore, after being told about this risk, the relevant information should be given by the doctor also in writing. This may help during contacts with other doctors consulted for confirmation of the received verdict.

Illustrated brochures containing basic information about this little known disease (such as those issued by some national tuberous sclerosis associations, e.g. in Britain[7] and the United States[6]) are also very helpful.

Thus, the general rule should be to inform patients and make sure that they are fully aware of the facts. However, sometimes such knowledge can adversely affect the patient or his/her relatives; it can lower self esteem, and cause unnecessary suffering. Consequently, all cases should be treated individually. If information about the nature of the disease has no practical importance for prevention or for treatment of the patient, then it does not seem absolutely necessary to disclose that a person is affected with an abortive form of the disease.

We have seen a seriously affected girl whose mother, a highly educated professional, was convinced that her daughter was a sporadic case of TS. However, detailed examination showed that the mother had an abortive form of the disease. She was, at the time, over 50 years of age, and had no other close relatives. There was thus no increased risk of recurrence of TS in the family. We considered that conveying information that the mother was affected could only bring harm to herself and to her family. Therefore, we decided not to reveal this information to her.

Thus, while observing general principles of genetic counselling, it should never be forgotten that the well-being of patients and their families are of primary importance. The knowledgeable information which can be provided may be helpful, but also harmful, and therefore it should be treated with great caution.

BIBLIOGRAPHY

1 Borberg, A.: 1951, *Clinical and Genetic Investigations into Tuberous Sclerosis and Recklinghausen's Neurofibromatosis,* Opera ex Domo Biologiae Hereditariae Universitatis Hafniensis 23, E. Munsgaard, Copenhagen, 103–105.

2 Connor, J. M.: 1985, *Linkage studies in tuberous sclerosis,* Paper read at the Tuberous Sclerosis Symposium of the Tuberous Sclerosis Association of Great Britain, 18–20 September, Nottingham.

3 Fryer, A. E., Connor, J. M., Povey, S., Yates, J. R. W., Chalmers, A., Fraser, I., Yates, A. D. and Osborne, J. P.: 1987, *Evidence that the gene for tuberous sclerosis is on chromosome 9,* Lancet, *1,* 659.

4 Gomez, M. R.: 1985, *Overview of the art of tuberous sclerosis. Clinical, pathological and imaging aspects,* Paper read at the above-mentioned symposium.

5 Lindenbaum, R.: 1985, *Prevalence of tuberous sclerosis,* Paper read at the above-mentioned symposium.

6 National Tuberous Sclerosis Association: 1985 *Tuberous Sclerosis. An Illustrated Brochure for Physicians, Publication No. 001/85.*

7 Wilson, J.: no date, *Tuberous Sclerosis,* Tuberous Sclerosis Association of Great Britain.

8 Zaremba, J.: 1968, *Tuberous sclerosis: A clinical and genetical investigation,* J. Ment. Defic. Res., *12,* 63.

S. KESSLER AND H. KESSLER

PSYCHOSOCIAL DEVELOPMENTS IN THE FIELD OF GENETIC COUNSELLING

In contrast to the exciting technological advances in prenatal diagnosis and in the capacity to diagnose and treat genetic disease, advances in the psychosocial arena of genetic counselling have been less dramatic and less obvious. The recent discoveries in molecular biology are truly revolutionary and applications of recombinant DNA technology are not only changing the map of the human genome but also the prospects for future medical practice. In comparison, those of us involved in the psychosocial side of genetic counselling have learned to have greater respect for the saying of Ecclesiastes – "There is nothing new under the sun". On the psychosocial side of genetic counselling, both gains and losses are evident. In this paper we take stock of the present status of the psychosocial dimensions of genetic counselling.

RESPONSE TO GENETIC DISEASE

The occurrence of genetic disease, or the threat of its occurrence constitutes one of the more severe psychological traumas. The desire, hope and dream for healthy offspring is universal. Even parents affected with achondroplasia (who prefer to have dwarfed children like themselves) want *healthy* dwarfed children. When the dream is frustrated by genetic disease, parents, sibs and other family members frequently show the various symptoms of post-traumatic stress reactions including depression and other grief reactions, diminished self-esteem, guilt cognition, disruptions of interpersonal relations and recurrent, intrusive waves of distressing thoughts and feelings.

SHAME AND GUILT

The occurrence of genetic disease robs parents of their dreams and undermines their wish for safety, stability and orderliness. It erodes the illusion that they have control over their lives and that the universe operates on the principle of fairness. Because childbirth is an event marked with major social significance, the occurrence of a birth defect frequently produces diminished self-esteem and marked reactions of shame. The latter include the various responses to exposure and narcissistic insult and are generally accompanied by strong affect and actions designed to undo the insult and to limit the injury to the self system.[15] Shame reactions often have major intrapsychic and interpersonal consequences and

189

E.K. Hicks and J.M. Berg (eds.), The Genetics of Mental Retardation, 189–199
© 1988 by Kluwer Academic Publishers.

sensitive, empathic interventions may have long-lasting implications for the counsellees and their relationship with the professional. As parents consider and review the possible reasons why the defect occurred, guilt reactions become prominent. The scientific explanations offered by professionals as an attempt to alter guilt cognition are frequently based on stochastic processes and hence require more abstract levels of thinking and evaluation than the persons may be capable of, given the confusion and distress they may be experiencing. The person is often asked to replace a seemingly coherent explanation of events based on self responsibility with one based on processes over which he/she has no control. Unless one has gained sufficient emotional distance from the initial trauma or the counsellor is particularly persuasive, the tradeoff may not seem reasonable to the person. Thus genetic counsellors often find that irrational and improbable associations between antecedent actions (or lack of actions) and the occurrence of a birth defect are tenaciously held despite argument and evidence to the contrary. The usual strategies of alleviating guilt employed in genetic counselling are extremely limited unless carried out with due attention to psychodynamic issues.

COPING WITH GENETIC DISEASE AND ITS THREAT

The desire to prevent birth defects and assure healthy offspring leads to the consideration and use of prenatal diagnostic technology. This faces parents with fundamental psychosocial issues. Some individuals have never considered the possibilty that they might produce a less-than-perfect child and they are distressed to learn about their vulnerability in this regard. Others find reassurance that genetic technology increases their control over the health of their progeny and provides a sense of mastery over the fear that they might give birth to a defective child. Also, parents are confronted with the need to make some important choices – to use or not use the available technology; to continue or terminate a pregnancy should the results of prenatal testing indicate an abnormal fetus. They must choose and live with the consequences of their choices. At stake are one's self-image as a sexual being and adult; one's competency in conceiving and bearing normal children, thus participating in a basic part of the life cycle; one's deeply felt beliefs regarding life, and its value relative to other important values.

Most parents recover from the psychological trauma of birth and genetic defects, "bloody but unbowed". However, psychological scars remain and often other life events, particularly other losses, reopen these wounds. When the pertinent coping skills, or timely interpersonal support and encouragement, are lacking, less than optimal personal and/or interpersonal readjustments may occur. Some individuals and relationships never recover from such traumas. For others, the occurrence of a birth defect resurrects past traumas and losses and provides an opportunity, albeit painful, for further psychological integration and mastery.

Attempts to understand how individuals cope with such major life stresses as the birth of a child with a genetic disorder have invariably turned to the grief model developed by Lindemann and others.[17,6] This model has been widely discussed in the popular literature and by the mass media. Lay persons appear as informed as professionals and their language is punctuated with key phrases and concepts. However, this often becomes one other way of intellectualising and distancing oneself from one's feelings. Also, norms of behavior have been established which leave persons feeling deviant when they find themselves responding in ways differing from those dictated by the model.[a]

PSYCHOLOGICAL IMPLICATIONS OF GENETIC COUNSELLING

Because genetic counsellors often see individuals and couples in the early phases of their coping with loss and/or the diagnosis of a genetic disorder, their psychosocial interventions may have an important bearing on how families mobilize and direct their coping skills to deal effectively with the problem. Genetic counselling in such circumstances is akin to preventive mental health and the counsellor's interventions or lack of them may have a long-term impact on the way families adjust. Obviously, the counsellor's beliefs regarding his/her professional role will influence the stance taken toward counsellees under these circumstances. Some counsellors believe that their role is simply to supply information. Others take a more active role in exploring how information is understood and in determining the effectiveness of the counsellees' psychosocial context. Still others take an even more active position by helping counsellees reach decisions and/or by teaching them more adaptive coping skills when such instruction appears necessary.

Many parents come to a genetic counselling clinic with enormous ambivalence and trepidation. On the one hand, they want to find out why things went wrong, what they might do to redress the situation and how they might prevent a future recurrence of the problem. Information gathering is the crucial step in the coping process; it both enlightens the person and re-establishes the sense (or illusion) of control.[14,25] Simultaneously, they neither want to be accused or blamed for the problem–and the fear is that they will–nor do they want to be hurt further by the information the genetic counsellor has to offer. Thus, counsellees often adopt avoidant, evasive and other self-protective tactics to prevent expected further harm; they both want to know and not know bitter truths. This ambivalence leads to considerable distortion of perceptions and of information processing which, in turn, interferes with dispassionate or so-called rational decision making. Commonly, counsellees selectively attend to data consistent with their need to maintain hope that their injury or hurt might be limited and contained, whereas information inconsistent with that hope receives lower attentional priority. Sometimes, professionals label this tendency as denial.[b]

PSYCHOLOGICAL IMPLICATIONS OF PRENATAL DIAGNOSIS

Couples coming to use prenatal diagnostic technology face peculiar and novel psychological issues. To ascertain the chromosomal or other genetic status of the fetus in the early stages of a pregnancy may interfere with the usual processes of maternal-fetal attachment formation. The parents appear to defer such attachment until they are reassured that the fetus is normal.[3] At the same time, such procedures as ultrasound visualization of the fetus may tend to accelerate this attachment.[7,20] Parents who desire the pregnancy are desperate for reassurance that everything is going well, and will take any sign that no problem has been detected as strong evidence of normality even though the risk for undetectable defects remains. To participate in prenatal testing opens the possibility of detecting an abnormality, and persons who are informed that this is the case need to decide whether or not to continue the pregnancy. For many couples, pregnancy termination is a difficult choice which runs counter to deeply held beliefs. How genetic counsellors prepare couples for possible abnormalities, help them make their decisions, and position them psychologically for the decision's aftermath, are activities with long-term implications for the couple. Perforce, in carrying out these functions, genetic counsellors cannot avoid wearing the mantle of the psychotherapist.

Some genetic disorders and birth defects are incompatible with life and, at times, genetic counsellors need to help parents deal with the issue of death, particularly in the neonatal period. Should parents be encouraged to hold and attach to a newborn infant who is likely to die? Similar questions might be raised about stillborn infants. How can professionals assist parents under such conditions to minimize long-term psychological trauma and/or dysfunction? When couples choose to terminate a pregnancy after a genetic disorder is detected, should they be encouraged to undergo a procedure which involves labor and birth of the fetus or is a D&C procedure under anesthesia advisable? Some genetic counsellors believe that if a person does not experience the full impact of such loss, the working through and completion of grief reactions might be impeded. Others believe that the strengthening of defenses is in the long-term interests of clients. Unfortunately, this area is difficult to research and no data are available to test competing psychological theories in this regard.

DECISION MAKING

There is increasing awareness of the difficulties inherent in human decision making.[9] Considerable evidence suggests that the evaluation of probability involves cognitive processes which, even under the best of circumstances, are subject to biases and rules of mental operation which lead to distortions and either under- or overestimation of risk.[10] Even experienced statisticians are not immune from these biases and are likely to make errors in judgment. If such distortions occur under conditions of low emotionality, they are also likely in circumstances of great emotional distress and uncertainty, such as those promoted by the occurrence of a birth defect or by the need to reach decisions about reproduction under the threatening cloud of genetic disease. Also, because these decisions invariably involve complex ethical/moral and psychological issues around the theme of life–its quality, its giving and its taking– decision making is particularly difficult for parents and often requires professional assistance. The lamentable and not uncommon practice of merely providing information and sending counsellees home to reach a decision on their own assumes that counsellees have not only absorbed and understood the information but also have the requisite decision making skills to deal with complex issues and their implications. When this is the case, the practice may be merited. But, how many of us have these abilities and skills? Under the best of circumstances, humans are generally poor decision makers.

GENETIC EDUCATION

One aspect of genetic counselling which has received generous research attention is the degree of success in achieving educational goals. Evers-Kiebooms and van den Berghe[5] and others have reviewed the relevant studies. In sum, there is room for improvement. In a large-scale, carefully designed study, Sorenson and his colleagues[23] showed that maximum educational success in genetic counselling was obtained among persons already well informed about recurrence risks and diagnosis. About half of individuals poorly informed about risk and/or diagnosis prior to genetic counselling remained so following counselling.

How counsellees integrate the information they acquire and how they use such information in later decision making is not well understood. We are slowly discovering that we also know very little about the processes involved in evaluating risk and using the risk figures obtained in genetic counselling for reproductive decision making. In one study[18], it was reported that, after genetic counselling, counsellees transformed risk figures into qualitative binary form – one will either have or not have an affected child – in order to use risk estimates in decision processes. Unfortunately, since the study was carried out after decisions were reached, it is not known whether such simplifying strategies were actually used in decision making or were *ex post facto* rationalisations.

PERCEPTION OF RISKS AND PROBABILITIES

It is also beginning to be realised that probability in the form of a recurrence risk is part of the linguistic context of genetic counselling. This means that, because it is given orally, the risk figure is assessed as language rather than as mathematical symbolism. Work carried out in Berkeley in collaboration with Dr. Eleanor Levine of California State University at Hayward suggests that how recurrence risks are conveyed may influence their assessment (Kessler and Levine, in press).

We found that the cognitive strategies used to assess proportions appear to differ from those used to assess percentages. Thus most subjects tended to perceive a difference in the magnitude of risk between proportions (e.g., 1 out of 4) and their equivalent percentages (e.g., 25%) in full knowledge that the figures were mathematically equal. In general, when asked to role play a parent coming for genetic counselling, the overwhelming majority of subjects perceived proportions with denominators of 10 or less as being riskier than equivalent percentages. Subjects tended to attend to the language with which risk was conveyed rather than to the mathematical aspects of risk. Saying that one has a risk of '1 in 4' tended to promote a pattern of thought involving an attempt to image 4 persons (or representations of persons) in which 1 was differentiated from the others. Providing a recurrence risk in the form of a percentage led to a different cognitive strategy in which a comparison of numbers was made without an attempt to form images of persons. The former strategy personalised risk by promoting an identification with the differentiated image on the part of the subject. In so doing, there was a tendency to magnify the magnitude of the risk figure. Providing a risk figure in terms of percentages appeared to promote a more abstract and depersonalized train of thought in which the relative risk magnitude of percentages was magnified. These findings suggest that it may be possible to influence and manipulate how counsellees perceive the magnitude of recurrence risks, which in turn may effect their later reproductive decisions. If confirmed that such is the case, important ethical questions will (and should) be raised. Should the counsellor attempt to influence decisions in the direction favouring the reduction of disease incidence? Before embarking on this path, considerably more will need to be known about how the language of genetic counselling influences (or fails to influence) counsellees and their decision making processes.

THE ROLE OF LANGUAGE

The language of counselling and psychotherapy is a subject under intense scrutiny in contemporary clinical psychology. Language is the means of organising our experience and understanding of the world.[26] Language is also the tool for effecting important changes of the person's understanding of reality and thus of his/her expectations and behavior.[2,16] With their more biological orientations, most genetic counsellors have not yet been exposed to these developments.[c] Very little attention has been given in the literature to the language of genetic counselling, and as little to the techniques of counselling. Beyond the most general of principles, no standards of practice or of professional conduct have been formulated to guide practitioners in their work. Perhaps even more important, there are virtually no data on what transpires in genetic counselling sessions. Thus some basic issues are in considerable doubt. It is not known, for instance, what is the degree of variation of counselling for similar problems in different centers or the degree of consistency of individual counsellors dealing with the same or differing problems. Although many counsellors have stated that they use nondirective approaches, no data exist to determine what they mean and how they achieve nondirectiveness. The genetic counselling session is a veritable black box.

TRANSCRIPTS OF COUNSELLING FOR EVALUATION AND RESEARCH

Several years ago, one of us[13] published a transcript of a genetic counselling session in the hope that others would follow suit, thus opening up the process of genetic counselling to more public examination and study. We retain faith that genetic counsellors will realise that this may be one approach to increase the effectiveness of counselling and that professional growth is being impeded by the absence of information that would allow the modification and modulation of counselling methods and procedures and serve as a basis for teaching genetic counselling skills.

PSYCHOLOGICAL DIMENSIONS OF GENETIC COUNSELLING

The psychological nature of the genetic counselling encounter is widely recognized. Increasing attention is being given to the possibility that, in many cases, the major work of the genetic counsellor is in the psychosocial realm.[11] This includes dealing with the counsellees' perceptions, thoughts and feelings, describing and discussing options, assisting in decision making, identifying areas of potential personal and interpersonal difficulty, helping them find appropriate resources for further consultation and follow-up, relieving feelings of guilt and battered self-esteem and, when necessary, teaching adaptive coping skills. It has become increasingly evident that genetic counselling deals with human behavior and that, although the focus is on genetic disease and reproduction, the work of the genetic counsellor is akin to that of the psychotherapist in affecting and influencing attitudes, beliefs, feelings, thoughts and actions. One cannot affect these psychological parameters in the limited arena of genetic disease and reproduction without influencing also other areas of the person's functioning.[12] The goal of genetic counselling might be thought of as an attempt to influence human behavior. The process of educating persons about their genetic makeup is an indirect way of discussing important psychological issues, such as relationships with significant others, identification processes, self-worth, repsonsibilitiy, autonomy, competence, and one's self- image as an adult, sex partner or parent. Educating counsellees is a way of influencing their attitudes, beliefs and behavior in these domains. The issue facing genetic counsellors is to clarify and define the goal or direction this influence should take. Should it focus on reproductive behaviors or those involving or associated with self-esteem, more adaptive functioning, and so on?

The genetic counselling literature of the last few years shows a clear reduction of the antipathy toward psychological issues. Several important books have been published in which the psychological nature of genetic counselling has received a prominent place. Examples are the volumes of Applebaum and Firestein[1], Emery and Pullen[4], Hsia and his associates[8], Riccardi and Kurtz[21], and Schild and Black[22], in addition to articles in both American and European journals. There are also some disturbing signs. With the increasing utilisation of genetic services, interest in efficiency has grown, particularly in assembly-line methods of counselling and educating. This is troubling for two reasons. First, if anything, genetic information is becoming more complex than ever before and this means that clients will need more rather than less human contact in order to understand, integrate and use genetic information for health and reproductive purposes. Automated educational approaches might be helpful in educating but are less useful in helping clients evaluate information, subjectively assess various options and reach personally relevant decisions. Secondly, assembly-line methods are out of step with the growing awareness in medical education and practice for the greater humanisation of the doctor-patient relationship. The over-emphasis on research and scholarship in medical training over the past two decades, rather than on the development of communication and interpersonal skills, has led, among other things, to increasing consumer dissatisfaction, increased noncompliance and professional ennui. Whereas in other fields of medical practice attempts are being made to reverse this maladaptive trend, medical genetics many be moving in the opposite direction.

COUNSELLORS, CLIENTS, AND SOCIETY

Many fundamental problems in genetic counselling persist. How can the inevitable conflicts between societal interest in preventing and reducing the social burden of disease be reconciled with individual autonomy in reproductive decisions? Whose interests does the genetic counsellor represent – societal ones, institutional ones or those of individual clients? What is the aim of genetic education? Is it neutral and value-free or is it a subtle form of behavior control?[19] How does the counsellor balance institutional needs and requirements, when providing information, and clients' needs to obtain simple and personally relevant information? How and under what circumstances should the counsellor be directive or nondirective? How can counselling neutrality and/or nondirectiveness be best achieved in the context of human interactions?

As genetic technology advances, new ethical, legal and psychosocial issues emerge. The development of RFLP linkage tests for the diagnosis of such adult onset diseases as Huntington disease (HD) promises to pose major dilemmas not only for the at- risk persons involved but for professionals and society at large as well.[24] If individuals participate in predictive testing as young adults and discover that they carry the gene for HD, how will they cope with this information? The absence of a cure or effective treatment for the disease may so demoralize some individuals that psychiatric dysfunction or suicide may result. Who might be liable under such circumstances? How can the livelihood and quality of life of persons found to carry the HD gene be protected? Should the new technology be used for prenatal diagnosis and should a fetus found to be a HD gene carrier be aborted even though many years of productivity and good health are likely prior to onset of the symptoms? Should minors be tested? How will parents and children deal with the information obtained from testing?

In California and elsewhere maternal serum alphafetoprotein programs have been initiated. As in other screening programs, unforeseen problems have emerged. Low alphafetoprotein values, among other implications, may indicate the presence of a Down syndrome fetus. Understandably, this would cause enormous unanticipated anguish. Since such screening was set up to detect neural tube defects the counselling provided prior to testing often has not prepared parents adequately for other sets of problems and counselling centers often have had difficulties in providing needed psychosocial services and follow-up. High alphafetoprotein values may indicate the presence of a neural tube defect and sometimes parents are overwhelmed by conflicting advice and standards of behavior in such circumstances. For example, some members of voluntary organisations dealing with spina bifida children may encourage parents to continue the pregnancy of such a fetus, whereas others might counsel having an abortion. Presumably, a major part of these differences in opinion arise out of parents' own experiences with a spina bifida child.

Increasing sophistication in the use and interpretation of fetal ultrasound visualisation in the prenatal period will enable detection of relatively minor conditions such as cleft palate which usually is repairable postnatally. Do such conditions merit the consideration (or encouragement) of abortion? In a similar vein, what societal standards of health and quality of life are likely to emerge as predictive tests are developed for such conditions as schizophrenia and other mental disorders, heart disease, and diabetes? Many of the old problems, such as the priority of individual vs social rights, will still remain. The patterns established today to deal with these difficult issues may set the tone for the future.

NOTES

[a] The traditional grief model leaves much to be desired. First, it is oversimplistic and does not account for individual variations in grieving behaviors. Secondly, it too strongly emphasizes a step-wise progression through stages of coping or grieving, whereas almost invariably actual responses show inconsistencies and simultaneous use of strategies from differing stages. Thirdly, the labels attached to the different stages explain the process from the professional's viewpoint not from that of the person undergoing stress. Thus, frequently behaviors are labeled negatively (e.g., denial, resistance) because they inconvenience the professional, whereas from the client's viewpoint they may be highly adaptive. Fourthly, the model is based on individual functioning, whereas, commonly, behavior is modulated and shaped in large part by interactions with others. Thus, counsellors often do not see the subtle differences in coping when a couple appears to be 'in-step', or the similarities when they appear to be 'out of step', in their grief or coping.

[b] The denial process needs to be better understood because it frequently puts a wall between counsellors and counsellees and reduces effective communication. When the parents of a new born child with Down syndrome cannot see the stigmata so plain to the professional, usually they are said to be denying reality. An alternate explanation is that lacking the professional's extensive

experience in dealing with birth defects, they do not have the skill to recognize the subtle differences between their baby and others. Also, the term 'denial' is applied to describe many other diverse circumstances, including difficulties in recalling diagnostic and other information, the minimization of the severity of a problem, having children in the face of genetic risk, seeking a second medical opinion and so on. It is doubtful that the cognitive/affective and decisional processes underlying these differing behaviors can be accommodated by a single psychological process, denial.

c A major area of interest in contemporary social psychology is the elucidation of the determining influence and power of implicit or indirect communication in human interactions. Metaphorical approaches, embedded perscriptive statements and other indirect ways of communicating are means of bypassing the person's usual defenses and providing persuasive suggestions as to actions the person might take. Psychotherapists have begun to realize the power of language and of indirect modes of communication in helping clients develop alternate, more adaptive, ways of thinking about themselves, their experiences and the world in general.

BIBLIOGRAPHY

1 Applebaum, E. G. and Firestein, S. K.:1983 *A Genetic Counseling Casebook*, The Free Press, New York.

2 Bandler, R. and Grinder, J.: 1975, *The Structure of Magic*, Science and Behavior Books, Palo Alto.

3 Beeson, D. and Golbus, M. S.: 1985, *Decision making: Whether or not to have prenatal diagnosis and abortion for X-linked conditions*, Amer. J. Med. Genet., *20*, 107.

4 Emery, A. E. H. and Pullen, I.: 1984, *Psychological Aspects of Genetic Counselling*, Academic Press, London.

5 Evers-Kiebooms, G. and van den Berghe, H.: 1979, *Impact of genetic counselling: A review of published follow-up studies*, Clin. Genet., *15*, 465.

6 Falek, A.: 1979, *Use of the coping process to achieve psychological homeostasis in genetic conditions*, In, H. A. Lubs and F. de la Cruz, eds., Genetic Counseling, Raven Press, New York, 179–191.

7 Fletcher, J. D. and Evans, M. I.: 1983, *Maternal bonding in early fetal ultrasound examinations*, New Engl. J. Med., *308*, 392.

8 Hsia, Y. E., Hirschhorn, K., Silverberg, R. L. and Godmilow, L.:1979, *Counseling in Genetics*, Alan R. Liss, New York.

9 Janis, I. L. and Mann, L.: 1977 *Decision Making*, The Free Press, New York.

10 Kahneman, D., Slovic, P. and Tversky, A.: 1982 *Judgment Under Uncertainty: Heuristics and Biases*, Cambridge University Press, Cambridge.

11 Kessler, S.: 1979, *Genetic Counseling: Psychological Dimensions*, Academic Press, New York.

12 Kessler, S.: 1979, *The genetic counselor as psychotherapist*, In, A. M. Capron, M. Lappe, R. F. Murray, Jr., T. M. Powledge, S. B. Twiss and D. Bergsma, eds., Genetic Counseling: Facts, Values, and Norms, Alan R. Liss, New York, 187–200.

13 Kessler, S.: 1981, *Psychological aspects of genetic counseling*, Amer. J. Med. Genet., *8*, 137.

14 Kessler, S.: 1984, *Psychologic responses to stresses in genetic disease*, In, J. O. Weiss, B. A. Bernhardt and N. W. Paul, eds., Genetic Disorders and Birth Defects in Families and Society: Toward Interdisciplinary Understanding. Birth Defects, Orig. Art. Srs. 20 (4), 114.

15 Kessler, S., Kessler, H. and Ward, P.: 1984, *Psychological aspects of genetic counseling. III. Management of guilt and shame*, Amer. J. Med. Genet., *17*, 673.

16 King, M., Novik, L. and Citrenbaum, C.: 1982, *Irresistible Communication*, W. B. Saunders, Philadelphia.

17 Lindemann, E.: 1944, *Symptomatology and management of acute grief*, Amer. J. Psychiat., *101*, 141.

18 Lippman-Hand, A. and Fraser, F. C.: 1979, *Genetic counseling the postcounseling period: I. Parents; perceptions of uncertainty*, Amer. J. Med. Genet., *4* 51.

19 London, P.: 1969, *Behavior Control*, Harper and Row, New York.

20 Milne, L. S. and Rich, O. J.: 1981, *Cognitive and affective aspects of the responses of pregnant women to sonography*, Mat. Child Nurs. J., *10*, 15.

21 Riccardi, V. M. and Kurtz, S. M.: 1983, *Communication and Counseling in Health Care*, C. C. Thomas, Springfield, Ill.
22 Schild, S. and Black, R. B.: 1984, *Social Work and Genetics*, The Haworth Press, New York.
23 Sorenson, J. R., Swazey, J. P. and Scotch, N. A.: 1981, *Reproductive Pasts, Reproductive Futures: Genetic Counseling and its Effectiveness*, Alan R. Liss, New York.
24 Wexler, N. S., Conneally, P. M., Housman, D. and Gusella, J. F.: 1985, *A DNA polymorphism for Huntington's disease marks the future*, Arch. Neurol., *42*, 20.
25 White, R. W.: 1974, *Strategies of adaptation: An attempt at systematic description*, In, G. V. Coelho, D. A. Hamburg and J. E. Adams, eds., Coping and Adaptation, Basic Books, New York, 47–68.
26 Whorf, B. L.: 1956, *Language, Thought and Reality*, M.I.T. Press, Cambridge, Mass.

DISCUSSION: GENETIC COUNSELLING

Who should provide counselling:

There was considerable debate as to what professional background (e.g. medicine, biology) was most suited for genetic counselling. It was generally agreed, however, that some formal training in genetics was imperative in effecting competent counselling. Moreover, it was felt that the most effective genetic counselling requires a team of specialists, including a medical geneticist, a social worker, and a psychologist.

Follow-up procedures:

The need for improved and more extensive follow-up of those counselled was discussed at some length. Of particular concern were the difficulties encountered in making sure that information conveyed was understood and remembered by clients. The following suggestions were made for improving follow-up procedures:

a. provide the family doctor (or referring physician) with all relevant information so that follow-up can be carried out at the local level.

b. send follow-up letters, or even tape recordings, of counselling sessions to the prospective parents and/or their families.

Sterilisation of the retarded:

Although some mentally retarded persons are sterile because of associated abnormalities, many are able to have children. The latter raises concerns in both the legal and medical professions in many countries. In Denmark, for example, sterilisation of the retarded, in general, is not permitted, with anticonception measures or abortion being the only alternatives. A rather similar situation exists in at least some regions of Canada. In the U.S., the parents of a retarded child can seek legal aid in effecting the sterilisation of their handicapped child, but permission is at the discretion of the judge on a case by case basis. In Hungary, sterilisation is prohibited for males and only exceptionally permitted for females. In Sweden, birth control is used as an alternative to sterilisation.

Screening for mental retardation:

Screening here implies full diagnostic evaluation, including prenatal and family history, dysmorphology evaluation, and relevant laboratory testing. Photographs

E.K. Hicks and J.M. Berg (eds.), The Genetics of Mental Retardation, 201–204
© 1988 *by Kluwer Academic Publishers.*

of the patient or certain relatives can also be useful, though this usually requires written consent, access to medical records, and perhaps even disclosure of medical data to relatives. This raises ethical issues surrounding non-disclosure. It can also prove difficult to convince institutional staff, parents, and others that the manifestations in a given child may relate to an abnormality which was present prenatally. In Denmark, for example, it is unusual for child psychiatrists to have their patients extensively tested to determine a possible physical cause for behavioural deviations.

Compounding these problems are the diagnostic difficulties inherent in identifying a syndrome on the basis of the presence or absence of specific features. Additionally, these features can vary from patient to patient, and can change with the age of the patient. Moreover, once labelled, a lack of interest often ensues in the biology (e.g., cytogenetic, biochemical) of a given syndrome. Consequently, new names may be given to variations of the same condition.

Genetic counselling and uncertainty:

What, and how much, should those seeking counselling be told? Should they be made aware of all relevant factors regardless of the degree of certainty, or should certain information be withheld? Some of the major questions raised with regard to these issues included:

a. Should the counsellor provide those counselled only with information they request, or also with that which they do not request?

b. Should the counsellor determine whether or not to withhold information, on a case by case evaluation, by judging what seems in the best interests of the persons counselled?

c. How is the counsellor to evaluate the understanding and coping capability of those counselled in terms of how they may interpret, or misinterpret, information given them? If a conflict exists between full disclosure and judgement of the capability of coping with the information, does the genetic counsellor become a therapist, and when does the counselling of such persons cease? It was agreed that this constituted a problem, and that a team counselling effort could help to mitigate it by involving follow- up counsellors.

d. Should uncertain preliminary diagnostic information be conveyed when subsequent testing might negate the earlier evaluation, or should information be withheld until all relevant tests have been conducted, and the degree of uncertainty thus hopefully minimized?

e. If manifestations of abnormality will not appear until later life, should the person counselled be forewarned of this, even though such information could easily result in serious psychological trauma for both the individual concerned and other family members?

f. If the information a counsellor has is not relevant to physical well-being, but could be psychologically disruptive, should that information be withheld?

g. What is a genetic counsellor to do when the issue is one of non-paternity? Many of the participants indicated that, where this is an issue, they would, at the request of the mother, usually opt to withhold from others information concerning non- paternity. The ethical/moral position on this issue, according to one of the ethicists present, was that the first priority is, and should remain, the genetic problem. Paternity disclosure decisions should ultimately rest with the mother, whether this involves contacting the actual father for testing, or informing her husband that he is not the father. The first priority of the counsellor should be to provide assistance to the mother.

Illustrating these important issues, regarding uncertainty of diagnosis and nondisclosure, was the discussion concerned with tuberous sclerosis. Individuals affected with this condition may be gravely or very minimally afflicted and the extent and degree of abnormality may not be apparent for some time.

Other problems associated with this disorder can include the identification of a parent as a carrier of the gene concerned, and the appropriate procedures to follow when a parent, who is affected but not obviously so, has an affected child. This latter situation engendered considerable debate, as it provided an example of compassionate non-disclosure. One specific reference was to a 50-year-old mother who was a carrier of the condition, but not obviously affected clinically, and had had an affected child. In this case, the physician did not deem it necessary to inform the mother that she was a carrier of the gene, as she could not have more children, and as her knowing that she was a carrier could aggravate her already unstable psychological status.

This raised the question of the ethics of non-disclosure of test results. The ethical-moral position advanced was that intentional non-disclosure should be exercised only when the disclosure of information might harm the person concerned. Moreover, an obligation to truth does not necessarily imply an absolute obligation to always disclose every detail. A direct question concerning test results must be answered truthfully, but where such questions do not arise, and the experience of the counsellor dictates that non-disclosure is in the best interests of both patient and family, then non-disclosure can be morally justified. It was stressed that non-disclosure must be evaluated on a case by case basis, and that rigid regulations in that regard would be inappropriate. Thus, it was generally agreed that non- disclosure was usually undesirable, but that in some instances it was justifiable.

A Hungarian programme:

The positive and negative aspects of a Hungarian programme for reproductive planning were debated at some length. The major points at issue included:

a. The directive aspect of the programme, in which pregnancy was discouraged in light of potentially severe, or non-treatable, medical outcomes.

b. Many participants in the discussion felt that this programme of optimal planned conception too closely resembles some eugenics programmes of the

1930s. It was countered that the present programme is a voluntary, not a mandatory, one.

c. It was suggested that relevant health education programmes in secondary schools and through the media would be more productive than a nationally controlled and operated programme. It was countered that neither time nor funding permitted the introduction of the former programmes, although it was agreed that they would be preferable. It was also countered that the present programme was developed on the incentive of a private institution, rather than by the State.

d. One problem with such a programme, it was pointed out, is its presentation to the public as a prevention programme for congenital malformations.

One participant predicted that, as overall potential for testing for carriers increases, there would be an increase in genetic counselling prior to a first pregnancy.

NOTES ON CONTRIBUTORS

Michael Baraitser, B.Sc., F.R.C.P., is a Consultant in Clinical Genetics, Hospital for Sick Children, London.

Joseph M. Berg, M.B.B.Ch., M.Sc., F.R.C. Psych. F.C.C.M.G., is Professor of Psychiatry and of Medical Genetics, University of Toronto; and Director of Genetic Services and Biomedical Research, Surrey Place Centre, Toronto, Canada.

Bruno Brambati, M.D., is Director of the the Perinatal Unit at the First Institute of Obstetrics and Gynecology, 'L. Mangiagalli' Clinic, University of Milan, Italy.

Andrew Czeizel, M.D., is Head of the Department of Human Genetics and Teratology, W.H.O. Collaborating Centre for the Community Control of Hereditary Diseases, National Institute of Hygiene, Budapest, Hungary.

G. R. Dunstan, M.A., Hon. D.D., Hon. L.L.D., F.S.A., is a Clerk in Holy Orders, Professor Emeritus of Moral and Social Theology at the University of London, and an Honorary Research Fellow at the University of Exeter.

Annalise Dupont, M.D., D.Sc., is a Paediatrician and Head of the Institute of Psychiatric Demography, Psychiatric Hospital, Risskov, Denmark

Karl-Henrik Gustavson, M.D., is Professor and Chairman of the Department of Clinical Genetics, University Hospital, Uppsala, Sweden.

Malcom A. Ferguson-Smith, F.R.C
Marie E.Ferguson-Smith, B.A., Dip. RCPath., is a Principle Scientist directing prenatal and tissue culture services for the West of Scotland, Department of Medical Genetics, Yorkhill Hospitals, Glasgow.

John C. Fletcher, Ph.D., is Chief of the the Bioethics Program, Warren G. Manguson Clinical Centre, National Institutes of Health, Bethesda, Maryland.

Jean-Pierre Fryns, Ph.D., Professor and Head of the Clinical Genetics Unit, Centre for Human Genetics, University of Leuven, Belgium.

E.K. Hicks and J.M. Berg (eds.), The Genetics of Mental Retardation, 205–207
© *1988 by Kluwer Academic Publishers.*

Arvid Heiberg, M.D., D.Sc., is Medical Director of the Frambu Health Centre, Siggerud, Norway.

Esther K. Hicks, M.A., Drs., Ph.D., is Head of the Science Research Department, Netherlands Universities' Joint Social Research Centre (SISWO), Amsterdam, The Netherlands.

Seymour Kessler, Ph.D., is Associate Clinical Professor, Department of Padiatrics, University of California, San Francisco.

Hilda Kessler, Ph.D., is a Clinical Psychologist in private practice in Berkeley, California.

Margareta Mikkelsen, M.D., D.Sc., is Head of the Department of Medical Genetics, J.F. Kennedy Institute, Glostrup/Copenhagen, Denmark.

M. F. Niermeijer, M.D., Ph.D., is Professor in the Department of Clinical Genetics, Erasmus University, and University Hospital, Dijkzigt, Rotterdam, The Netherlands.

Dick Smithells is Professor of Paediatrics and Child Health at the University of Leeds, England.

Louise J. Thomassen-Brepols, M.D., is a Psychologist at the Stichting Klinische Genetica, Regio Rotterdam, Erasmus University, Rotterdam, The Netherlands.

M. A. M. de Wachter, Ph.D., is Director of the Institute for Bioethics, Masstricht, The Netherlands.

Ignacy Wald, M.D., D.Sc., is Professor and Head of the Department of Genetics, Institute of Psychiatry and Neurology, Warsaw, Poland.

Dorothy P. Warburton, Ph.D., is Assoc. Professor of Clinical Paediatrics (in Genetics and Development) at the College of Physicians and Surgeons, Columbia University, New York, New York.

Udo Wilken, Ph.D., is Professor of Special Education and Social Ethics at the Fachhochschule, Hildesheim, West Germany.

Ishwar C. Verma, F.R.C.P. (London), F.A.A.P. (U.S.A.), F.A.M.S. (India), is Associate Professor of Paediatrics and Head of the Genetics Unit, Department of Paediatrics, All India Institute of Medical Sciences, New Delhi.

Jacek Zaremba, M.D., Ph.D., is Senior Lectuer in the Department of Genetics, Institute of Psychiatry and Neurology, Warsaw, Poland.

A. Zwinger, M.C., D.Sc., is Head of the Centre for Prenatal Diagnosis of Genetic Disorders, at the Institute for the Care of Mother and Child, Prague, Czechoslovakia.